INFESTED

BROOKE BOREL

INFE

THE UNIVERSITY OF CHICAGO PRESS · CHICAGO AND LONDON

INFESTED

HOW THE BED BUG
INFILTRATED
OUR BEDROOMS AND
TOOK OVER THE
WORLD

BROOKE BOREL is a science writer and journalist. She is a contributing editor to *Popular Science*, where she authors the blog *Our Modern Plague*.

The University of Chicago Press, Chicago 60637
The University of Chicago Press, Ltd., London
© 2015 by Brooke Borel
All rights reserved. Published 2015.
Printed in the United States of America

24 23 22 21 20 19 18 17 16 15   1 2 3 4 5

ISBN-13: 978-0-226-04193-3 (cloth)
ISBN-13: 978-0-226-04209-1 (e-book)
DOI: 10.7208/chicago/9780226042091.001.0001

"Party Rock Anthem" lyrics printed with permission of Peter Schroeder, David Listenbee, Stefan Gordy, and Skyler Gordy. Published by Nu 80's Music LLC, Redfoo LLC, and ESKAYWHY Publishing. Administered by Kobalt Music Publishing America, Inc.

LIBRARY OF CONGRESS CATALOGING-IN-PUBLICATION DATA

Borel, Brooke, author.
   Infested: how the bed bug infiltrated our bedrooms and took over the world / Brooke Borel.
      pages; cm
   Includes bibliographical references and index.
   ISBN 978-0-226-04193-3 (cloth : alk. paper) — ISBN 978-0-226-04209-1 (e-book)
1. Bedbugs—History. I. Title.
   QL523.C6B67 2015
   595.7'54—dc23   2014035967

♾ This paper meets the requirements of ANSI/NISO Z39.48-1992 (Permanence of Paper).

For Mike,
who endured years of bed bug stories
at every dinner party

Slovenly man has carried this abominable bug to all parts
of the civilized world.

LELAND OSSIAN HOWARD, 1902, CHIEF
OF THE DIVISION OF ENTOMOLOGY, US
DEPARTMENT OF AGRICULTURE

The June bug's got the golden wing,
The lightning bug the flame,
The bed bug's got no wing at all,
But he gets there just the same.

The pumpkin bug's got a pumpkin smell,
The squash bug smells the worst,
But the perfume of that old bed bug,
Is enough to make you burst.

When that bed bug comes down to my house,
I takes my walking cane,
Go get a pot and scald him hot!
Goodbye, Miss Liza Jane!

NURSERY RHYME, CIRCA 1900S

Now do not raise your hands in amazement, I beseech you, nor turn away, with uplifted eyes and disdainfully curled lips, at my want of refinement, at my utter disregard of taste. Do not even marvel at my presumption in presenting to your notice the humble, despised, loathed objects discussed in this paper. Cast aside, for this one and only time, I pray you, prejudice and old associations, and believe the sad, inglorious tales hanging around them false, or at least much exaggerated. Fancy me a second Cinderella, dwelling in by-corners, searching and sifting the debris of your houses, and spying out hidden secrets of Unwelcome Guests, who will visit you whether you like their company or not.

CHARLOTTE TAYLOR, "UNWELCOME GUESTS," *HARPER'S NEW MONTHLY MAGAZINE*, DECEMBER 1860

CONTENTS

PROLOGUE
Mysterious Bites

Late in the summer of 2004 in New York City, I watched my doctor take a ballpoint pen and trace the perimeter of a welt of unknown origins on my right leg, which had spread from my shin around to the center of my calf. He told me: "If the red extends beyond this line, go back to the emergency room."

We had blamed the welts, for there were others, on everything from spiders to mosquitoes to ticks. I had been tested for Lyme disease; fed a round of antibiotics to extinguish an infection that I had likely picked up from scratching too much; and prescribed a pack of steroids to quell the painful swelling of my allergic reaction. The Lyme test was negative. And although my pills made my red streaks and lumps disappear, the relief was not to last. Despite five visits from an exterminator over the course of two months, the bites always returned, and always after I'd slept in my bed. Whatever was lurking in my room at night was still there.

When the bites began, I had been living in New York for two years and was temporarily working at a law firm while I applied to graduate schools and studied for the relevant entrance exams. Any spare money I had went to traveling; I stayed in hostels and other budget accommodations to stretch my meager funds. When I was at home, my Hell's Kitchen apartment became a stopover for friends visiting from across the country and overseas, as well as a regular gathering spot for new friends from around the city. The welts cast a shadow on this socializing. They were a mystery I needed to solve in order to reclaim my living space.

The answer didn't come from the experts I'd turned to, but from my dad back in Kansas. A pathologist who specializes in conditions of the skin, he may have come across a reference to my problem in a dermatology journal, although now, a decade later, the details are murky. Regardless, one night on the phone as I recounted my woes, he asked: "Do you think you have bed bugs?"

"Bed bugs? Are you crazy? That's not even a real thing," I said.

But after I hung up, I turned to my computer to look up these imaginary beasts and learned that they actually *are* real. Not only that, the common bed bug is a distinct species with its own name, *Cimex lectularius*—one cousin in a family of more than a hundred. Once I worked through the cognitive dissonance caused by my belief that the bed bug was nursery rhyme make-believe and the discovery that it was an animal with a proper Linnaean name, I hired a new pest controller and spent weeks dragging my clothes and bedding down to my building's basement laundry room, where I washed them in scalding water and dried them on high. I tore through my closet, flipped my mattress again and again to inspect each corner, and emptied my dresser drawers. I bagged my decontaminated clothing and vacuumed for hours, tracing each of the fine cracks in my rental's parquet floor.

All this digging turned up a single bed bug. Its origin I will never know. Maybe it was a visit to Ireland, where I stayed in a different countryside inn each night, or a long weekend in Las Vegas. I took both trips in the month before my first bite appeared, and bed bugs can spread on the luggage of weary travelers. Or perhaps the bug snuck in from a neighbor, although my apartment manager swore there were no other insect problems in the building.

Eventually I put my closet back together and tucked my clothes and sheets into their proper places. The following spring I saw an article in the *New Yorker* detailing the bed bug travails of three young women in their East Village apartment, and I felt less alone. The stigma eventually gave way to humor, and my experience became nothing but a story I would tell at a bar. Gradually, I even began having houseguests and parties again.

But five years later I'd repeat the entire process in Brooklyn. Twice. Both times in the same summer in two different apartments. First, the bed bugs came to my now-husband's apartment in Greenpoint. We pinpointed those to his stay in a hostel in San Diego; a friend who had shared a room with him had also taken bed bugs home to his own apartment in San Francisco. Then bed bugs showed up in my apartment ten blocks away, although it's still a matter of debate whether they came from my husband's old place or a used futon that my new roommate had bought off of the Internet. During that long summer of bagging, vacuuming, and struggling to put expensive protective covers on our mattresses and box springs, I started seeing the bugs on the news, on blogs, and on sitcoms. They would come up in casual conversation. CBS declared 2010 "The Year of the Bed Bug." What was going on?

By the time of this media explosion, I was working as a freelance science journalist. I pitched a short article on bed bugs to an editor at *Popular Science*, which gave me an edge I didn't have five years before—a legitimate excuse to call entomologists and ask as many questions about bed bugs as I wanted. During these interviews, I learned that the bed bug, which most people thought we had conquered after World War II with the debut of modern pesticides, was in truth an ancient pest. Bed bugs lived with our ancestors at least since the pharaohs ruled Egypt and possibly much longer, stretching back to the Pleistocene, before modern humans even existed, when the bugs may have made a living on the blood of both bats and our close relatives, who occasionally sought shelter in the same caves. From there, the bugs shadowed us throughout history, solidifying our bond once we moved to permanent camps and cities, and conquering the world as tiny bloodsucking colonists.

This new information made it feel increasingly odd that I had grown up with no knowledge of the pest. For my generation to be oblivious of the bed bug, I realized, was as strange as the thought of future children not knowing the cockroach, the ant, or the fly. And I began to understand what I think is the most intriguing aspect of the bed bug's story: that they are back today isn't a fluke. It is a return to normal, an ecological homeostasis.

And, man, are they back. In 2004, the year I fed my first bed bug, there were 82 confirmed bed bug violations reported to New York City's Department of Housing Preservation and Development. By 2010, the number had risen to 4,808. New York is not alone. Today the insects are in every state in the United States. According to a 2013 survey, more than 99 percent of American exterminators have treated for bed bugs over the previous year, up from 95 percent in 2010, 25 percent a decade before, and 11 percent before that. The United States is not unique in its widespread bed bugs. In Australia the common bed bug and its cousin, the tropical bed bug, collectively jumped a shocking 4,500 percent between 2000 and 2006. Similar trends are seen worldwide, and although some argue that the scourge's spread is slowing down—which, in fact, it is not—the cosmopolitan pest continues to spread into smaller and smaller cities and towns like a fast-growing mold.

After I wrote that initial news story, I wanted to learn more. And so to find out how the bed bug returned with such a vengeance and how everyone else was handling it, I set out on an adventure that turned into a book project that, unexpectedly, led me to scientists nourishing bed bugs on artificial blood feeders and even on their own arms and legs; to the front row of an Off-Off-Broadway bed bug rock opera; to housing projects in Ohio and Virginia; to historical accounts of entomologists traveling the world to track the bugs; to Roma communities in eastern Slovakia; and to bat roosts in the Czech Republic. Throughout it all, our intense reaction to the small pests surprised me again and again.

So, too, did the bed bug.

Let me show you.

INFESTED

ONE

CRYPTIC INSECT
Meet the Bed Bug

Picture a bedroom. Maybe it's yours. Maybe the bedding is clean and crisp, a laundry-fresh comforter is tucked around the mattress, and your clothes are hidden away, neatly folded in your dresser or hanging in your closet. Maybe, instead, the sheets are twisted, the blankets askew, and your jeans from yesterday are on the floor next to the hamper. It doesn't matter. Somewhere in that bedroom, small secretive bugs may have squeezed into a crack or hole imperceptible to your clumsy eyes: the joint of the bed frame, the head of a screw in the back of your nightstand, or perhaps a fold in the lining of the suitcase that is still sitting, unpacked, in the corner. The bugs are reddish brown and flat, and are most comfortable in these tight spaces, where they spend most of their time waiting. Waiting for you.

These are bed bugs. Their knack for concealment is why entomologists sometimes call them cryptic insects, although the uninitiated often think, incorrectly, that they've never seen a bed bug not because it is good at hiding but because it is invisible to the human eye. Somehow, although our history with this ancient pest stretches back many millennia, its brief sixty-year absence from a large swath of the world shrank our impressions of its physicality to microscopic dimensions. It became both an imaginary and an invisible threat. This made the bed bug's return as a real animal that takes up space in the world—our world, our beds—all the more unsettling. But despite the bed bug's ability to hide and to seem invisible, it is not. Some who have seen one say it resembles a drop of blood with legs. Others offer less gruesome analogies: an

Cimex lectularius, the common bed bug. Credit: Richard Naylor, CimexStore.co.uk.

adult bed bug is the size and shape of a lentil or maybe an apple seed. Whatever the comparison, the insect is a physical being. You can cradle it in the palm of your hand, look into its tiny eyes, and watch it march across your mattress.

While a bed bug's life may seem secret to us, it carries on the same basic routines as any other animal: it eats, seeks shelter, and has sex. For a bed bug, food is always blood. It hunts down each blood meal, as entomologists call it, every few days to a week and almost exclusively at night. From its hiding place in the bed-frame joint or the nightstand screw, it senses the carbon dioxide from your breath, the heat from your body, and, perhaps, some of the hundreds of other chemicals regularly emitted from your skin. It ventures out, scurrying across the floor, up the bed legs, and across the sheets. When the bed bug finds you, it grips your skin with clawed feet and unfolds its mouth—a long tube called a proboscis, also called a beak—to probe the flesh, seeking the best place to bite. Within the beak are the bug's upper and lower mouthparts—the maxillae and mandibles, respectively—each divided into right and left sides. When the bed bug is ready to penetrate the skin, the toothed mandibles lead the way, snipping through like scissors to make a path for the maxillae, which follow. Once inside, the mouthparts restlessly seek a blood vessel. Unlike some insects that guzzle pooled blood, the bed bug is a bloodsucker and takes its meal from blood circulating inside a living thing. Assisted by the difference between the high pressure of the blood vessel and the low pressure of its empty body, it fills like a water balloon attached to a spigot.

To find the perfect spot where the blood flow isn't too fast or too slow, the bed bug's mouth performs extraordinary acrobatics, sometimes bending more than ninety degrees as it explores the flesh. Once the bug settles on a vessel, it injects saliva packed with a cocktail of forty-six proteins. Some are anticoagulants to prevent clotting, for a blood clot would be deadly as a half-chewed hunk of steak lodged in your throat. There isn't much room to play with. The bed bug's mouth is just eight micrometers in diameter—thinner than a strand of silk, but just wide enough, as a human red blood cell is seven and a half micrometers across. Other bed bug saliva proteins act as vasodilators, which widen the blood vessels, or prevent hemostasis, which keep the blood flowing; still others have antibacterial properties or help with lubrication. Like other blood-feeding insects, the bed bug may also numb its host with proteins that act as anesthetics to help avoid detection, although no one has scientifically proved this.

An adult bed bug's bite lasts around eight minutes, during which its flat body plumps to double or even triple its original size. Young bed bugs, called nymphs, require less blood, although they need to feed at each of their five stages in order to grow. If they don't, they remain in arrested development indefinitely, or at least until they starve to death. After a bed bug feeds, it concentrates the protein-rich red blood cells, squeezing the rest of the meal— mainly a liquid blood component called sera—out of its rear mid-bite. These drops and, later, the fully digested blood meal, fall to the bed sheets and dry as black stains, a telltale bed bug mark. Sometimes, too, bed bugs leave a signature as a line of bites along a person's body, a result of several bugs biting at the juncture where the skin meets the bed sheets. ("Like pigs to the trough," as I've heard one medical entomologist describe it.)

After feeding, an adult bed bug skitters back to its bed-frame joint or screw head or suitcase, or wherever else it has made its home, at speeds of up to four feet per minute. Nymphs move considerably slower. Both find their way with specialized receptors on their fine antennae and, perhaps, in their feet, which detect chemicals called pheromones that help guide insects' social behavior, oozing from other bed bugs back at the refuge. These are called aggregation pheromones for the fact that they encourage the bugs to group together. (All bed bugs also emit alarm pheromones in times of danger to warn others away, and females may also use chemical signals to help nymphs find their first meal.) Once a bed bug has tracked down the aggregation pheromones and it is safe in its hiding spot, it snuggles in with anywhere from five to dozens of others, including both nymphs and adults. They pack in tight amongst their own eggs, cast skins, and shit, giving off a musty, fruity odor that was described in 1936 by an entomologist as an "obnoxious sweetness."

After a meal, bed bugs often engage in rough-and-tumble sex, in part because a satiated female is sluggish and her plump body

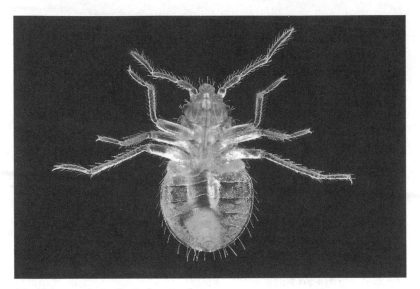

A bed bug nymph. Credit: Graham Snodgrass.

makes her easy to mount. (Once, at a bar, I overheard an entomologist call the bugs "chubby chasers" for this fact.) Bed bugs are part of a club of just a handful of invertebrate classes that mate by an unusual practice called traumatic insemination, and they hold the title for the most highly adapted form, as well as the most studied. It goes like this: The male bed bug climbs onto his lover's back, his head resting on the left side of her pronotum, the outside of the first segment of the thorax that is roughly equivalent to her neck. He grasps her with the claws of his feet and tucks his abdomen so that it curves around her body, holding her in a violent embrace. At the tip of his abdomen is a hypodermic appendage called a lanceolate paramere, which is essentially the bed bug's penis. He swiftly stabs the female's underside and ejaculates into her body cavity. It's more like a shanking than a romantic coupling.

To counteract the male's stabs, the female bed bug has evolved a unique protective organ called a spermalege. On the outside, it appears as a small notch on the right side of her segmented abdomen, which physically guides the paramere to the correct spot—the male may stab anywhere, but the female subtly directs his aim. Inside the spermalege, a mass of immune cells called hemocytes, analogous to white blood cells, protects the female from bacteria that coat the paramere and helps heal her love wounds. The sperm, which have their own anti-microbial properties that may shield them from bacteria and help further protect the female, make their way into the female's circulatory system and ultimately collect in bags attached to her ovaries. There, she stores the sperm, using just a bit at a time to stretch her fertilization period to between five and six weeks. This is a particularly handy strategy if she is swept away, alone, to a new home.

Once fertilized, the female may lay five eggs a week, which can add up to several hundred over her lifetime, although just as with any animal, this varies from one individual bug to another, impacted by her circumstances, such as access to food and the temperature of her home. However many eggs she lays, she exudes a gummy substance that cements them together and to the surface where they are laid. ("Looks like mini caviar," a gruff Brooklyn exterminator once told me, although I later decided they are more like tiny grains of rice.) The female bed bug lays these eggs from a genital tract that is perfectly functional and could be used for a more traditional mating style. Yet still, the male stabs, and more often than necessary. A female bed bug needs to mate just once for every four blood meals in order to produce the highest possible number of eggs. Males typically initiate sex twenty to twenty-five times that often. If a population of bed bugs has too many males, they may stab the females to death in their overenthusiasm and leave them no time to heal from their wounds. Or so some scientists claim.

Entomologists have been trying to understand the bed bug's strange mating ritual for nearly a century. The answer may lie in an evolutionary biology concept called sexual conflict, where one sex of a species evolves features that increase its chances of breeding, but at the other sex's expense. Their partners co-evolve strategies to counteract these unwanted advances. This sexual arms race is relatively common. Fruit flies battle over how long the male's sperm can survive in the female's reproductive tract. Muscovy duck males have alarmingly long curlicue penises, while the females have labyrinthine vaginas that twist in the opposite direction to turn away unwanted suitors. As for the bed bug, the male's stabbing ways may have evolved so that he could better compete

Traumatic insemination. Credit: Richard Naylor, CimexStore.co.uk.

with his rivals; by stabbing his paramere directly into the female's body, closer to her ovaries, he might get his sperm to the target a little faster than a beau that chose the old-fashioned route. Other research suggests that the last sperm into the female's sperm bags are the first to make it to the ovaries, which means the last male to mate with a female may get to cut the paternity line. This could explain why males mate at such mad frequency. Yet another possibility is that the male's approach evolved in response to female adaptations that tried to thwart his moves. Regardless of the cause, and unfortunately for the female bed bug who lacks good resistance against over-mating, it appears that she is currently losing this particular battle of the sexes.

All this activity describes just the hidden world of the common bed bug, *Cimex lectularius*, which is the species that has reasserted itself in temperate climates worldwide over the past fifteen years (in this book, "bed bug" will generally refer to *C. lectularius*). Around a hundred related species live out shadowy parallel lives, mainly in the nests of birds or the roosts of bats rather than your hypothetical bedroom, where they feed on pigeons and chickens and sparrows and mouse-eared bats. Collectively, these bugs are called Cimicidae (pronounced *sigh-MISS-uh-dee*), or cimicids.

Cimicids belong to the Linnaean order Hemiptera, or the "true bugs," which is why entomologists spell out "bed bugs" as two words (by contrast, a house fly is two words because it's a true fly, while a butterfly, one word, is not). All Hemiptera have sucking mouthparts in order to puncture the outer layers of the flora or fauna on which they feed and to drain the fluids inside, whether it's the sap in a leaf or the juice in another insect or, for cimicids, the blood of a more complex animal. The true bugs' name comes from *hemi-* for "half" and *-ptera* for "wing"; they are named so for the fact that their wings form a hardened shell at the top, where they attach to the body, and a thinner membrane toward the tips. The wings of many Hemiptera form a signature X pattern when folded across their back. But cimicids have only the stunted nubs of rudimentary wings, possibly because they have adopted the simplest eating strategy an animal can have, which is to sit and wait for food to arrive. The bugs have no reason to fly because we always return to our beds, just as birds do to their nests and bats to their roosts. Wings would also burden cimicids, tangling in feathers and fur during a meal.

Other than the common bed bug, which lives primarily in temperate regions around the world, only two other cimicid species regularly feed on human blood. One is the tropical bed bug, *Cimex hemipterus*, the common bed bug's counterpart in the tropics. Like the common bed bug, this species recently resurged across the world, particularly in Asia, India, and Australia, the latter of which

which is overrun with the common bed bug in the cooler southern regions and the tropical bed bug in the warmer north. The other is *Leptocimex boueti*, which feeds on bats and people in West Africa.

To the untrained eye, bed bugs, bat bugs, and the various bird bugs all look pretty much the same. They have wide-set eyes and short antennae. Their oval bodies are segmented, giving them a striped appearance, and bristle with stiff hairs, as do their three pairs of legs. Adults are dark brown and their color shifts to mahogany when they feed. Nymphs are white with eerie red eyes when they wriggle from their eggs. As they mature, they turn the color of straw and then gradually darken to brown. But while they may look the same to most of us, each has characteristics befitting life in its own unique niche.

• • •

When I first sought to uncover the secret life of the bed bug, I spent two hours on a Saturday morning on the phone with an entomologist who was studying the insect. It was the only day she had time to talk; bed bugs had taken over her life, and she worked all hours during the week studying the insect and fielding panicked calls from people with infestations.

During our interview, the entomologist told me about a book called *Monograph of Cimicidae*. It was published in 1966, and although some of its contents are outdated, it remains the heftiest bed bug book in existence, phone-book thick with 585 dense pages describing the seventy-four bed bug species that were known at the time it was written. When I sought a copy in 2011, it was relatively easy to find because the Entomological Society of America had issued a reprint the year before, at the height of the bed bug panic. At $74, it worked out to a dollar per species. The book started me on my much longer bed bug journey, remained a constant companion even as I dug through other papers and texts, and would pop up in places I did not expect.

Monograph of Cimicidae was written by a man named Robert Leslie Usinger, an entomologist who worked and taught at the University of California, Berkeley, in the forties through the sixties. He was also the most significant contributor to bed bug research in the last century. When Usinger wrote his tome, and somewhat still today, entomologists identified cimicid species by slight variations in the bugs' legs, bodies, and other prominent features that are indistinguishable to my naked eye. But when I flipped through the monograph for the first time, the variety from one species to another was dizzying. The magnified views of Usinger's clean ink drawings, as well as those by a few colleagues, spanned more than 150 pages. The species had their own portraits, which revealed that some bugs are long and thin while the others are short and fat; some are balding, others hirsute, and the hair of each grows in a unique pattern; antennae may stick out ninety degrees from the side of the face, or angle toward the front, or bend back to point at the insect's rear end; legs are long and spindly or short and squat or something in between; and abdomens range from thin and pointy to a near-perfect circle. Detailed figures compared antennae of many species at once, illustrating distinct lengths and shapes and constellations of hair, or lay out the differences between a series of legs, some of which are mottled and others uniformly shaded. Still other drawings compared bed bug genitalia. One series showed female bat bugs and bird bugs and bed bugs, the varying shapes and positions of their varying spermaleges delicately outlined, and another laid out the array of bed bug penises, all of which crooked to the left but differed in length, girth, curvature, and pointiness.

Usinger wrote several pages on each cimicid species, but he dedicated most of his monograph to *Cimex lectularius*. Bats and birds, after all, may be tormented by bites, but they can't intellectualize what plagues them (or, for that matter, hire an exterminator). People are a different story. We spill the most ink for the animals that spill the most of our blood. The enigmatic common bed bug is no exception.

• • •

How did this strange insect come to be? And how is it now so widespread? In a way, we created the modern bed bug: it evolved to live on us and to follow us. We became an efficient vehicle to spread it around the world. As such, its long story and uprising is intricately intertwined with our own history, and understanding its path helps illuminate ours.

Usinger suggests, and most experts today agree, that the bed bug got its start in caves somewhere along the Mediterranean seaboard tens of thousands to hundreds of thousands of years ago in

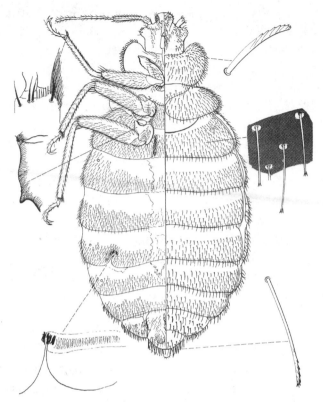

Drawing of a female common bed bug from *Monograph of Cimicidae*. Credit: Robert Usinger, courtesy of the Entomological Society of America.

what is now considered the Middle East. Bats likely lived in those caves, as they still do today, and they were host to parasitic bugs. Hypothetically, our ancestors—or perhaps close kin such as the Neanderthals, with whom our early relatives interacted and sometimes even had sex—sought shelter in these caves. When they did, some of the bat bugs took notice. Here was a new potential source of food. Temporary parasitic insects such as these are uniquely adapted to their host, especially when they live a restricted life with access only to a certain food source. Their mouthparts and legs, for example, are shaped to deal specifically with the skin and blood of whatever animal they feed on. These early bat-feeding bugs that were able to also bite our ancestors would have had characteristics allowing them to feed on an entirely new mammal with a strikingly different biology and lifestyle.

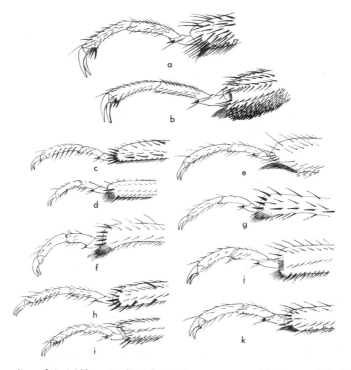

Comparison of cimicid legs. Credit: Robert Usinger, courtesy of the Entomological Society of America.

The shift from one host to another, and the subsequent altered version of the bug that would eventually emerge, was likely a messy process, and the precise moment when it began is unknown. But a simplified version of the story is this. Compared to the bugs living off of bats, the bed bug would evolve wider and longer mouthparts to accommodate for our ancestor's larger red blood cells and thicker skin. The newer bug developed less hair to make it easier to climb over our ancestor's smoother bodies. Its legs lengthened, morphing from short and strong, which helped to grasp a bat's furry body, to long and quick, which made it easier to run from slapping hands. Its circadian rhythm shifted, too, so it could feed at night, rather than during the day when bats roost; even today, the bug changes its feeding schedule to match its host's sleep cycle. The new bugs passed these favorable traits to their offspring. And as time went on, people began to live in more clustered homes in camps and villages, and the relationship with the bed bug grew

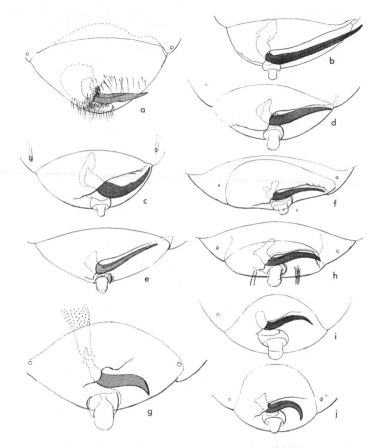

Comparison of cimicid penises. Credit: Robert Usinger, courtesy of the Entomological Society of America.

stronger. The bug thrived in increasingly condensed dwellings, its reproduction and spread boosted by the heat of our hearth-warmed homes. As early civilizations expanded interaction with one another through trade and travel and moved from smaller villages to cities, the bed bug did, too.

The earliest hard evidence of our coexistence with the bed bug comes from Tell el-Amarna, an Egyptian archaeological site that lies 170 miles south of Cairo and dates between about 1352 and 1336 BCE. Ancient Egyptians occupied the city that used to exist there, Akhetaten, for around a quarter of a century, during the reigns of the pharaohs Akhenaten and Smenkhkare. The city fell just before the start of the rule of Tutankhamun, or King Tut. The region's hot, dry climate preserved the flesh of insects including fleas, several food pests, and what have been identified as bed bugs, which were excavated in what may have been the former sleeping chambers of the el-Amarna tomb builders and guards. Bed bugs were still present in the region more than a thousand years later, appearing in a third-century BCE Egyptian papyrus that described a spell to keep them away. And they were likely still there beyond that. By at least the ninth century, the bugs were nearby in what is now Iraq, according to the Arabic scholar al-Jahiz, who wrote: "They feed on warm blood and have a crazy preference for man. They have no protection, so one easily sees them. In Egypt and similar lands, they are very prevalent."

Early historical documents do not differentiate between a common bed bug and a tropical one, particularly in the regions where the species may have overlapped. But some version of the bug lurks in the texts of three of the world's oldest living religions that sprung from Middle Eastern lands: Christianity, Judaism, and Islam. Christian references to the bugs appear in *The Apocryphal New Testament*, which describes, among other things, legends of second-century apostles. In the book's story of John, bugs attack the apostle at night. He tells them: "I say to you, you bugs, be considerate; leave your home for this night and go rest in a place which is far away from the servants of God!" And they did until morning. For their good behavior, John told them they could return to their homes, and so they "hastened from the door to the bed, ran up the legs into the joints and disappeared." The passage doesn't specifically call the insects "bed bugs," but they would indeed live in the joints of a bed (although they'd never be so well-behaved, God's servants or not).

In Judaism, the Talmud explicitly speaks of bed bugs at least twice. The most promising reference dates between the fourth and sixth centuries and recalls Mishnah Niddah, the classical rabbinical code that explains the rules on women's purity and menstruation. This particular Mishnah states that if a woman has blood on her bedsheets or clothes, she is considered impure and has to refrain from sex until seven days have passed and she has taken a purifying ritual bath. But if she can show that the blood is from some-

A preserved bed bug found at an Egyptian archaeological dig dating to between 1352 and 1336 BCE. Credit: Eva Panagiotakopulu.

thing else—from a cut, or from slaughtering animals earlier in the day, or from a crushed bed bug—she's still pure. (Today in Yiddish, the bugs are called *vantsn*, which translates literally as "bed bug," loosely as "small disgusting creatures," and forms the root of *vantzel*, which is used as an insult against people who are short.) Finally, some Islamic academics think there may be a bed bug reference in the Sahih al-Bukhari version of the Hadith, a record of the Prophet Muhammad written in the ninth century. In this Hadith, Muhammad says: "When anyone of you goes to bed, he should dust it off thrice with the edge of his garment . . ."—which the scholars interpret as clearing the space of bed bugs in preparation for sleep.

As best as we can trace the bed bug's spread, it advanced from what is now the Middle East and North Africa through Europe and Asia. While it isn't known for certain how the bugs made the trip across the Mediterranean Sea, a possible scenario is that they hitched rides on the ships that crisscrossed the waters during the Bronze Age sea trade. By at least 423 BCE, the bugs were in Greece, according to descriptions by the Greek playwright Aristophanes;

later the bugs were also mentioned by Aristotle and Dioscorides. In Greece they earned the name *kóris*, which some etymologists suggest is the root of the word *coriander*, possibly because the spice's crushed seeds give off a sickly sweet smell similar to that of a bed bug infestation. The ancient Greeks considered the bug both an undesirable pest and a homeopathic treatment for a long list of ailments: they hung hare or stag feet from their beds to ward off the bugs, yet ate them with meat and beans to treat fevers, with beans alone to cure snakebites, and with wine or vinegar to ward off leeches. They also crushed the bugs into balms to treat urinary pain or to regrow plucked eyelashes, and made the insects' fluids into depilatories.

By 77 CE bed bugs landed in modern-day Italy, according to Pliny the Elder, a Roman scholar and encyclopedist; later they would appear in the poems of Horace. Romans gave the bed bugs the first half of their modern Latin name, *Cimex*, which translates as "bug." Linnaeus, that great classifier of life, extended the designation to *Cimex lectularius* nearly a thousand years later, which means "bug of the bed" or "bug of the couch." Like the Greeks, Pliny said the bugs could be used medicinally, suggesting an injection of burnt bed bug ashes and rose oil for earaches, although he questioned the wisdom of eating them with beans.

From the edges of the Mediterranean, bed bugs emanated farther east and north, their spread made easier by our increasingly congested cities and penchant for travel. They entered through large cities or seaports and were relatively uncommon farther inland at first, although they eventually spread throughout the countryside. By 600 they were in China, where the ancient Chinese mashed the bugs into a paste to treat external sores and eventually would name them *chòu chong*, or stinky bug, in Mandarin. In nearby Japan, the bugs were called *tokomushi*, floor bug; *tokojirami*, floor louse; and *Nankinmusi*, the Nanking bug—possibly named for the former capital of China, a city that the Japanese associated with diminutive and rare foreign curiosities.

By the eleventh century, the bugs were documented in Germany, where they would eventually be named *venerschen* (little venereal), *nachtkrabbler* (night crawler), and *tapetenflunder* (wallpaper flounder). Centuries later the German writer Johann Wolfgang von Goethe would describe the demon in his famous Faustian legend as "the lord of rats and eke of mice, of flies and bed bugs, frogs and lice." By the thirteenth century, bed bugs were in France, where they were named *punaise*, from *puer*, or "to stink." By 1583 they made it to England, where they were dubbed "bug," the earliest use of which allegedly specifically referred to the bed bug and may have originated from the Scottish or Welsh word for bogey, goblin, or ghost. Soon the bed bug was so common in Europe that the popular natural history *The Animal Kingdom*, first published in 1817 by the French zoologist Georges Cuvier, claimed that it was "too well known to need description," although the book dedicated a page of dense text to the domesticated dog.

From Europe, colonists unwittingly brought bed bugs to America aboard their ships—even on the *Mayflower*, according to folklore. Some Englishmen claimed it was the other way around and that the bed bugs snuck into England in timber shipments from the New World—an unlikely suggestion since bed bugs feed on people, not wood, and wouldn't have infested virgin lumber to begin with. By at least 1748, the bugs were common from the English colonies up into Canada, first appearing in gateway seaport towns and then working their way into other settlements through trade and travel.

As settlements grew, so did the bugs' territory, and eventually they would earn nicknames like "the redcoats" and "mahogany flats." They hitched rides into homes on used furniture, travelers' luggage, and maids' laundry baskets, spreading between adjoining houses and adjacent tenements. They were prominent in offices and theaters. Soon, like their hosts, bed bugs headed west on covered wagon caravans and, later, by railway to conquer the entire continent. One account of the rugged prairie life recalls "swarms of bed bugs" that could be scooped from the walls of sod houses and measured with a spoon. Eventually, Native Americans incorporated the pest into their languages. In Navajo the bed bug is *wósits'ílí*, and in Cherokee it is *galuisdi*. In Hopi the bugs, or perhaps a closely related species, are *pesets'ola*, which serves as the base for the verb *pesets'olmaqnuma* and means "to be hunting for bed bugs," a phrase used when someone is falling asleep. Bed bugs similarly infiltrated Australia when the first European settlers ar-

rived there in the late 1700s, according to the few existing historical accounts that mention the insects, and likely also snuck into
other temperate colonies across the world. These regions, too, have
their own bed bug histories.

By the early 1900s, the pest was so common in the United States
that it turned up everywhere from downtrodden alleys to fancy
hotels. There was a Bed Bug Hill in New Jersey and Bedbug, California, a mining town. Fake bed bugs were sold for around ten
cents a package in magazines such as *Popular Mechanics* and *Bill-
board*. The American realist painter Edward Hopper made his own
bed bug gag, painstakingly crafting a watercolor bed bug family,
cutting each member out, and then attaching them to the artist
Walter Tittle's pillow as a practical joke. Americans mixed bed
bugs in tinctures and other medical mixtures to treat fever and
chills associated with malaria, as well as constipation, coughs,
hemorrhoids, liver complaints, muscle contractions, skin ailments, seminal emissions, and frequent yawning. And in 1926 a
special agricultural bulletin from Michigan State University pro-
claimed that bed bugs had "been established all over America for
so long a time that no record of a 'bugless' America seems to exist."

From there the bed bug worked its way not only into daily life
but into our artistic expressions of it. The British medical entomologist James Ronald Busvine, whose experiments helped prove
the bed bug's hardy resistance to pesticides in the fifties and sixties, understood how a bloodsucking insect could spark creativity:
"They will provide literary curiosities as symbols of piety, of love,
of human insignificance. They are subjects of ribald verse, of quack
medicine and of morbid fascination."

Indeed, bed bugs were written about in the pages of Upton Sinclair, Sinclair Lewis, Langston Hughes, John Steinbeck, John Dos
Passos, William Burroughs, Allen Ginsberg, and Henry Miller, who
loved the bug's name so much that it appeared in seven of his most
famous novels. The bugs, too, were lamented in early blues songs,
from "Black Snake Moan" by Blind Lemon Jefferson to variations
of "Mean Old Bedbug Blues" by the likes of Lonnie Johnson, Furry
Lewis, and Bessie Smith. In 1936 the song crept into American
country music when Ernest Tubb, the Texas Troubadour, recorded
a version complete with guitar picking and yodeling. And in the

fifties and sixties in the Caribbean, popular calypso tunes about bed bugs took on a sexual theme. One song recorded separately by Mighty Spoiler and by Lord Invader contemplated reincarnation as a bed bug in order to bite women's buttocks. And in "Muriel and the Bug," recorded by Lord Kitchener, a bug seeks out Muriel's "treasure."

• • •

Back in your bedroom—the clean or messy one, whomever you are—let's say you have woken up to an itchy bite or a smear of dried blood on your sheets. Or perhaps you've noticed a spray of small flecks that look like mold or ground pepper stuck to the edges of your mattress. Maybe this has happened before, but you just now accepted that there is a pattern. You see that the bites are lined up in a row or that the blood spot has the remnants of a crushed insect leg. Or maybe, on closer inspection, what looks like ground pepper is in fact bed bug scat. You now realize that these are the signs of the nocturnal romping of bed bugs. Maybe you did notice before but didn't want to believe it. It doesn't matter. This time let's say your findings spark a bed bug hunt, and you rip off your sheets to find a small cluster of bed bugs huddled in the seam of your mattress inches from where your pillow normally rests. And so you do what people have been doing for hundreds or thousands of years when faced with bed bugs: you kill them, and by whatever means possible.

TWO

THE FALL

DDT and the Slaying of the Beast

In 1935 in a tidy lab in Basel, Switzerland, a young chemist named Paul Hermann Müller doggedly released into a large glass chamber group after group of blue bottle flies. Müller, a serious man with a furrowed brow and dark hair cut high and tight, spritzed each set of the iridescent flies with a different insecticide dissolved in ethyl alcohol or acetone. He had been working in the chemical industry since he was seventeen years old, first as a laboratory assistant and later, after studying chemistry at the University of Basel, as a research scientist. But the experiment with the flies was his first attempt at working with insecticides, having spent most of his career investigating vegetable dyes and natural leather tanners. He had a deep love for plants and nature, and his new goal was to chemically protect plants from insect attack. He had an ideal formula in his head, which would be toxic to many types of insects but not to mammals or plants, work quickly and adhere to surfaces for a long time, cost little to make, and have no unpleasant stink. But nicotine, phenothiazine, pyrethrum, rotenone, and thiocyanate weren't right. Neither were the more than three hundred other chemicals he tried. Undiscouraged, for he knew a good chemist was a patient one, his mantra became: "Now, more than ever, must I continue with the search."

A basic chemical shape flickered through his memories from various experiments he'd both read about and performed. He dug through scientific papers and uncovered a chemical close to what he wanted in a 1934 article, and synthesized it in his lab. He tinkered with the shape, studding the chemical with chlorine atoms here and there, and tested the resulting compounds on his flies. In September 1939, four years after he had started his experiment and

Chemical structure of DDT.

as World War II broke out across Europe, he mixed his best shape yet. He found the compound in an old thesis submitted by an Austrian student in 1873, which described concocting it but made no mention of its use for insect control. The compound was lovely and symmetrical.

When Müller sprayed his flies with the compound, they dropped dead. The chemical clung so firmly to the inside of the test chamber that it also killed the next round of flies, which hadn't been treated with anything. Before he could use the chamber again, Müller had to take it apart, sanitize it, and air it out for an entire month. He was on to something.

Müller wasn't actually interested in killing blue bottle flies. His eventual targets were agricultural pests including the Colorado potato beetle, an invasive species threatening Swiss farms. He tested his new compound on that beetle. It worked. He tested it on house flies and gnats, too. It worked. The patient chemist also learned that this compound fulfilled nearly all of the other items on his insecticide wish list: it appeared to be safe for mammals and plants, it was cheap and easy to make, it was virtually odorless and stable in both sunlight and in air, and it could be used both indoors and outside. Within a year, Müller's employer, J. R. Geigy Ltd., was granted a Swiss patent (today Geigy is part of the chemical giant Novartis). Soon the company had two products on

Paul Hermann Müller, the inventor of the insecticide DDT. Credit: Science Source.

the market in Switzerland: Neocid, a louse powder, and Gesarol, a general-use spray.

In the coming years, scientists would discover that the chemical wasn't only effective against plant- and waste-eating pests, but also against bloodsucking parasitic insects—a category notoriously difficult to control that includes malaria-carrying mosquitoes and typhus-carrying lice. Seventy years later, thumbing through Usinger's *Monograph of Cimicidae*, I would read that this chemical—dichloro-diphenyl-trichloroethane, or DDT—also "seemed the perfect answer to man's age-old problem" of bed bugs.

• • •

DDT was the first of many synthetic chlorinated hydrocarbon insecticides. "Hydrocarbon" refers to a compound made of carbon and hydrogen atoms. These are organic compounds, meaning that carbon forms their foundation. All life on this planet is carbon based, from bed bugs to humans; carbon is a major building block in key biological structures including proteins, fats, carbohydrates, and DNA, in part because it is one of the few elements that can form long chains to build myriad shapes. "Chlorinated" means that some of the hydrogen atoms have been plucked off of the carbon and replaced by the element chlorine. Each chlorine substitution changes the characteristic of the hydrocarbon, forming everything from chloroform to cleaning solutions to paint thinner. Chemists learned to add other atoms and groups of atoms to the hydrocarbons to form increasingly complex chains of molecules. DDT is one example, a central carbon atom bonded to one hydrogen atom; a carbon atom that is, in turn, attached to three chlorine atoms; and a pair of benzenes, or carbon and hydrogen atoms arranged in a ring, with a chlorine atom attached to each.

This lovely, symmetrical, dangerous configuration makes DDT insoluble in water but soluble in fat, including fatty body tissues. In its natural state, which is a fine white powder that resembles baking powder, the toxic chemical doesn't move easily through the skin of humans and other mammals. DDT's structure is also responsible for its long-lasting residues, which stay on surfaces for months or even years.

Straight DDT was never used as an insecticide. Instead, it was diluted into carriers, including oil-based liquids or fine powders, to be sprayed or painted or puffed anywhere an insect might walk through or land on. When an insect touches a DDT-laden surface, tiny crystals of the compound stick to its feet. Later, the pest might clean itself, wiping the pesticide on other parts of its body. Just how DDT kills isn't entirely clear, which is also true of many of the chemicals that we make and use as poisons or medicines—scientists know they work but are not quite sure how. The assumption for DDT is that the waxy outer layer of the insect's exoskeleton absorbs the pesticide, after which it works its way into the nervous system. There, it interferes with the transport of important ionized salts that are key to transmitting the nerve signals responsible for movement and thought, or at least an insect's equivalent to thought. In particular, DDT appears to affect parts of a neuron's membrane called sodium channels, which help pass the sodium ions from one side of the membrane to the other as a signal zips between nerve cells and through the body. The ions leak out of these channels, causing the neurons to continuously fire, which in turn causes the insect to convulse uncontrollably. Then it dies.

DDT marked the first time that we would successfully mix our own insecticidal concoctions, linking together chemical components in a lab rather than relying on noxious elements, such as arsenic and mercury, or naturally occurring botanical poisons, such as pyrethrum powder, which is made from crushed chrysanthemums. DDT's success would pave the way for a number of other chlorinated hydrocarbons including aldrin, chlordane, dieldrin, lindane, and methoxychlor.

• • •

The story of DDT is well known. The pesticide's fast and widespread use was the product of two perfectly timed phenomena during World War II. The war ushered in an era of Big Science, driven by massive and expensive government-funded research. In the years leading up to the war, relevant research and development ramped up in the world's superpower states, including Germany, the UK, and the Soviet Union. The United States was no exception. In 1941 Franklin D. Roosevelt signed Executive Order 8807 to establish the Office of Scientific Research and Development in order to aid national defense. During its tenure, the OSRD and several other government institutions would create the bazooka, a method to mass-produce penicillin, and the DUKW amphibious vehicles now used for sightseeing "Duck Tours" around the country. The war also prompted the development of the first programmable electronic computer, advances in radio technology, the first intercontinental ballistic missiles, and, most notoriously, the first nuclear weapons. DDT and the insecticides that would soon follow benefited from a similar wartime interest and investment in new chemical technologies. In 1944 *Time* magazine called the insecticide "one of the great scientific discoveries of World War II," comparable to "Lister's discovery of antiseptics," and *Newsweek* ranked it in the top three medical discoveries from the era along with plasma and penicillin.

The second factor that aided in DDT's swift rise was that war presents a favorable breeding ground for infectious diseases, with thousands of humans writhing in combat and living in tight, often-squalid quarters. In most wars that came before World War II, armies suffered more casualties from disease than from combat, of which bacteriologist Hans Zinsser famously wrote: "Soldiers have rarely won wars. They more often mop up after the barrage of epidemics. And typhus, with its brothers and sisters,—plague, cholera, typhoid, dysentery,—has decided more campaigns than Caesar, Hannibal, Napoleon, and all the inspector generals of history." The pathogenic microbes that cause these diseases require a vector in order to spread from one host to another. For many, this vector takes form in a biting insect.

Early in World War II, it appeared that this tragic trend would continue. Two major threats included epidemic typhus, a contagious and debilitating disease caused by *Rickettsia* bacteria and mainly spread by body lice, and malaria, caused by a plasmodium parasite spread by bites from the females of certain species of mosquito in the genus *Anopheles*. Some medical interventions for both typhus and malaria existed but were limited. While Zinsser and others helped develop an effective typhus vaccine in the thirties, the use of broad-spectrum antibiotics against *Rickettsia* wouldn't be discovered until 1948. Synthetic anti-malarial drugs were initially developed in the late 1800s and were in even wider use by World War II. But getting vaccines and treatments to people in far-flung, war-torn locations was difficult if not impossible, and preventative measures such as draining the standing water in which mosquitoes breed or killing lice by washing bodies and clothes with soap and hot water weren't easy to do in wartime chaos. As long as infected insects survived, there was no way to totally prevent the outbreaks that spread as the bloodsuckers flew or crawled from person to person in jammed soldiers' quarters, POW camps, and prisons. Both pests ran rampant through troops on both sides of the conflict. (Bed bugs also harassed the troops, even at the highest ranks: in 1943 General George S. Patton complained more than once of bed-bug-ridden living quarters in letters to his wife, Beatrice, including the first time he wrote home after invading Sicily.)

Enter the war-technology machine. According to the historian David Kinkela's excellent account of DDT, in 1942 J. R. Geigy sent samples of Gesarol to its US subsidiary in New York, and not long after researchers both there and in an American government lab in Florida conducted the first studies showing the effectiveness of DDT against lice and mosquitoes. This was quite a feat considering the differences in the insects' feeding habits and general biology. *Anopheles* mosquitoes require a blood meal to reproduce, so they must live relatively close to humans. They also require water, spending the first three of their four life stages—egg, larva, pupa, and adult, respectively—in whatever water they can find, from irrigation ditches to a stagnant puddle in a rutted tire track. The mosquitoes may fly many miles to breed and to feed, which they do mostly at dusk or at night, depending on the species. Body lice, however, are wingless, and so they must stick close to their host at all times, eating, reproducing, and defecating in the seams of clothing and undergarments. For a pesticide to be used against both insects, it would need not only to be effective, but it would also need to be safe to use on the walls, ceilings, and pools of water where mosquitoes congregate and on lice-infested human bodies. DDT appeared to pass all tests. The researchers could spray it on any surface, water included, and when they doused people with it, it caused only the occasional rash. These safety tests were basic, giving no hint of long-term exposure, rushed in part because of the wartime pressure to wipe out disease.

Subsequent field tests in New York, New Hampshire, and Mexico, sometimes without consent, proved DDT's effectiveness outside of the lab, as did later tests elsewhere in the United States and in Trinidad, Egypt, and Brazil. By early 1943, the Brits and the Americans added it to their army supply lists. The insecticide arrived on battlefields just in time. There wasn't enough pyrethrum powder, a common insecticide of the era, to take care of the growing wartime insect problem, in part because Japan, an enemy of the United States and other Allied troops, controlled over 90 percent of the world's supply. Japan's ally, Germany, also received samples of Gesarol from Geigy (Switzerland, after all, was neutral) and submitted it to a battery of tests. But the Germans were skeptical of its benefits and wary of its potential health risks, and used it only to control malaria in German-occupied countries including Greece and Yugoslavia.

DDT got its big break later in 1943, after the Allies wrested North Africa and southern Italy from Benito Mussolini and found Naples and several smaller Italian villages in the throes of a potentially catastrophic typhus outbreak. It was the first major typhus problem in the region in about fifteen years, which made the locals especially vulnerable. An estimated 90 percent had lice. The region was in tumult from the sudden change of control from Axis to Allied forces. There was a disruption of communication and transportation that could have brought in food, soap, and other supplies; and bombings by both the Germans and the Allied forces drove people into hundreds of shared air-raid shelters carved into Naples's limestone hills. A typhus outbreak was inevitable.

By December 1943, the philanthropic Rockefeller Foundation's Health Commission launched a delousing program for more than 1.3 million people across southern Italy, including around a million in Naples. Using hand dusters in temporary stations set up in churches, schools, and railroad stations, they covered up to 100,000 people in DDT dust mixed with other delousing compounds every day. They powdered the homes of any known typhus patients, as well as those of immediate family. Within two months, the epidemic was under control. Later that year, DDT obliterated *Anopheles* mosquitoes in the region; and not long after, it was the insect killer of choice in war zones all over the world. DDT helped World War II become the first major war with fewer American deaths from disease than from battle injuries and among the earliest wars with lower disease deaths worldwide. The discovery of the chemical's insecticidal properties would earn Müller the Nobel Prize in Physiology or Medicine in 1948.

Just before the conflict ended, the US War Production Board, which had funneled materials into the wartime effort, lifted a ban on the use of domestic DDT. Soon the pesticide and other chlorinated hydrocarbons were wielded in a subtler war, taking place on farmland and in homes across the world. It was the battle against agricultural and domestic pests, including the bed bug. American businesses jumped at the opportunity, particularly those that had dominated military DDT contracts: DuPont, Hercules, Merck, and Monsanto. Following the production board's release, Gimbels department store paid for a page in the *New York Times* announcing the pesticide's arrival: "Released Yesterday! On Sale Tomorrow! Gimbels Works Fast!" In the following years, direct advertisements from Dow, DuPont, and others touted DDT's use on farms, gardens, and homes, while the US Department of Agriculture handed out brochures doing the same. One popular campaign for a product called Flit was illustrated by an up-and-coming cartoonist named Theodor Geisel, who would later become the children's book author Dr. Seuss. Another included a full-page color advertisement for Pennsalt Chemicals in a 1947 issue of *Time* magazine, which showed a dog, an apple, a housewife, a cow, a potato, and a chicken joyfully singing: "DDT is good for me-e-e!"

The powdered insecticides were soon available for public use, no permit or license required, sold in local pharmacies, hardware stores, and garden supply outlets for just twenty-five cents per pound. Housewives and home gardeners diluted the dust to 5 or 10 percent mixtures to spray liberally on beds, curtains, cribs, and lawns. Home decorators plastered rooms with DDT-impregnated wallpaper; there was even a Disney-themed print for the nursery with Mickey Mouse and friends frolicking on a background of flowers and balloons. Home owners painted their drains and screen doors with Pestroy and Certicide DDT varnish. Even dogs had their own dedicated DDT products. Cartoon mutts in an advertisement for Pulvex Flea Powder sang: "Fleas don't bother me, I'm dusted with Pulvex DDT," while cans of the DDT-powder RiDDiT-X were printed with the tagline: "You Lucky Dog." And a television ad from the fifties featured a woman spraying her Irish setter with an aerosol can of DDT and then rubbing the insecticide into the dog's lush red fur with her bare hand.

"DDT is good for me-e-e!" Credit: Oceana Wilson, courtesy of Crossett Library, Bennington College.

Over the three decades following the war, the United States would dump 675,000 tons of DDT across the country, dusting orchards by plane and directly spraying people and their belongings in an atomic-like bombing. Worldwide, the figure would grow to 2 million tons. With the surfaces and contents of millions of homes across the United States coated in a fine layer of insecticidal residue, the bed bug seemed to disappear.

DDT wasn't the only factor in the bed bug's demise. Prior to the war, Americans had cut down the number of bed bug populations, at least in middle-class and wealthy homes that could afford help with the laborious housekeeping that kept the bugs away or hire an exterminator savvy with a wide range of poisons. Across the Atlantic, Brits enjoyed a similar dip in the thirties, before DDT's insecticidal prowess was even discovered. Through toil and sweat, teams demolished old publicly funded housing and inspectors checked and rechecked homes for vermin. Other experts credit lower bed bug numbers to modern cleaning marvels such as the vacuum cleaner, the use of which surged in the postwar era. Still, while ripping up infested housing, vigorous inspections, and careful housekeeping may have helped control the bugs, none of these approaches were as simple, widespread, or effective as dousing a home in DDT.

• • •

Evolution follows a predictable pattern. Nature puts pressure on a species, whether in the form of food scarcity or predators or anything else that threatens survival and reproduction, and some individuals have a trait that helps them live on while others do not. Maybe, for example, the lucky organisms can access food that others can't, or they are better at evading animals that would eat them, and maybe this feature is written in their genes—a random mutation that makes them unique. The survivors have the opportunity to reproduce, passing whatever genetic characteristic spared them to their offspring, who similarly endure until that feature is widespread. This process is called natural selection because it is the environment that chooses who lives and who dies and shapes the attributes of a population, although this choice is without intention.

For most complex organisms, such as humans, some of these changes may happen quickly, but others take place over incredibly long stretches of time—hundreds of thousands or millions of years. In part, this is because vertebrates have longer gestation periods, compared to simpler creatures, meaning it takes longer for them to give birth to the next generation. They also give birth to fewer offspring, and it takes more time for those offspring to mature to reproductive age. Such large-scale change sometimes leads to populations of organisms with gene pools that have become so vastly different that they form new species. How quickly this happens depends, in part, on whether individuals from different populations can migrate back and forth and thus mix the gene pools; greater separation, possibly by physical barriers, prevents the groups from breeding and contributes to their divergence into new species, while more opportunities for migration leads to more gene flow.

But some organisms reproduce at far higher rates, giving birth to hundreds or thousands over shorter periods of time. This enables a new trait to evolve relatively quickly, especially in creatures such as bacteria or insects. Compare the time span of a human generation to that of the bed bug. A couple that met and married during World War II may have started their family during the post-war baby boom. Let's say they had five kids, who each grew up to have their own families. During the years that original couple produced two generations, bed bugs made hundreds of generations containing millions of more bed bugs.

Each generation was born in an environment doused with pesticide. The weakest bugs died, but some had characteristics, perhaps due to genetic mutations, that made them resistant to the pesticides, a synthetic environmental pressure even deadlier than a food shortage or a predator. These bed bugs survived the sprays and powders and varnishes, continued taking their nightly blood meals, performed their bizarre mating rituals, and spawned new generations of bed bugs with superior DDT-resistant genetic mutations; because bed bugs reproduce quickly, the resistant genes took over the gene pool. (This is similar to our current problem with antibiotic-resistant bacteria, where certain pathogens can withstand drugs, eventually leading to superbugs such as methicillin-resistant *Staphylococcus aureus*, or MRSA.)

DDT was hugely successful at wiping out most of the bed bugs, in part because the bugs had never experienced such a powerful pesticide and had no natural defense against it. While resistance spread quickly, our predecessors soon moved on to other pesticides, including organophosphates and carbamates, which were effective at first because they worked differently than DDT. (Organophosphates are still quite effective, although they are now illegal to use indoors in many countries.) Still, the DDT-resistant genetic mutations would remain in the DNA of small pockets of surviving bed bugs scattered around the world, ready to spring back under the right conditions. During that time, 76 million baby boomers grew up blissfully unaware of the bed bug, as did their younger brothers and sisters and their daughters and sons, knowing only that it was something that existed when their parents were young.

THREE

THE FORGOTTEN ERA
Out of Sight, Out of Mind

In 1965 at the Limited War Laboratory in Aberdeen, Maryland, army entomologists were testing bed bugs for combat. The Vietnam War had escalated on the other side of the world, and the Vietcong were putting up a more impressive fight than the Americans and their allies had expected. The enemy's familiarity with the jungle made it easy to ambush US soldiers; the army was thus desperate to flush the guerillas out. The Americans stripped the leafy jungle greens with Agent Orange and other defoliants so that the trees could no longer conceal an attack, and they trained German shepherds to sniff out hidden enemies sneaking through the ruined forest remains. But the researchers at Aberdeen thought that the bed bug might be a more versatile lookout. Compared to dogs, insects were easier to transport, required less care and attention, and needed no training.

Bed bugs are also naturally attracted to humans. The scientists wanted to exploit this tendency, whether in this species or a short list of other bloodsuckers, to see if they could detect the heat coming off of an enemy's body or the carbon dioxide from his breath. In addition to the bed bug, the contenders included an unnamed species of lice; *Xenopsylla cheopis*, the Oriental rat flea; *Amblyomma americanum*, the lone star tick; three mosquito species; and *Triatoma infestans*, the kissing bug and most infamous carrier of Chagas disease.

The scientists put each species through a series of tests to observe how the insects acted when a person was nearby and to see whether that action could be converted into a warning signal. Lice were ruled out early because their aimless crawls didn't change when a person was nearby. The fleas did notice the presence of a person, but they became so excited when they smelled a potential meal that they tapped like kernels of popcorn against a metal detection chamber and took too long to settle back down, which meant they were sensors that couldn't reset. The soft feet of the tick made no discernible noise even after the researchers hung weights from the arthropods' legs in hopes that the additional heft might audibly scrape across a detector's surface. In tests on one species of mosquito, the insects responded by probing a screened, skin-like membrane, thinking it was food whenever the researchers wafted in a human scent. A phonograph pickup, the same device that captures the vibration from the strum of an electric guitar and converts it to an electrical signal, connected to the membrane and converted the action of each wishful bite so it resonated like a plucked guitar string. And the kissing bug, a distant cousin of the bed bug, made a raucous noise with each of its footsteps, which was promising.

Both adult and nymph bed bugs sprung to attention when a meal was nearby, but only the younger bugs reacted strongly enough in initial tests to warrant the construction of a complex sensor. The researchers made one from a coiled spring of piano wire connected to a phono pickup. Just as the mosquitoes bit the fake skin in their detector, the bed bug nymphs shimmied across the piano wire, triggered it, and produced a sound. But when the wires, the bugs, and the pickup were put in a portable container—a small mesh envelope—the sound was too muffled to hear. The intrepid scientists built a chamber of fine steel wool and put the bugs and the detector inside. This helped, but the device still was not good enough to be useful on the battlefield. None of the other insect finalists that the army entomologists tested worked out, either, and the project was abandoned.

That the bed bug was included in the Aberdeen research in the sixties was unusual—as the bed bug faded from homes and our collective memory, it also became less common in the laboratory. In the decades leading up to World War II, scientists had mainly tried to understand the bugs' basic biology, whether they were a health threat, and how to kill them. Both during and directly after the war, the research slanted toward pest control through experiments using DDT and other poisons. By the late fifties, just

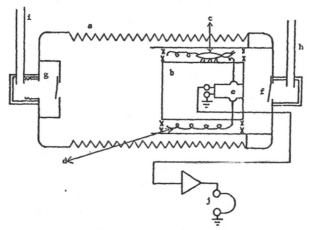

Figure 2. Diagram of insect ambush detector. a. Bellows pump.
b. Cylindrical plastic tube. c. Bug. d. Piano wire entangle-
ment. e. Phono pickup cartridge. f. Intake valve. g. Exhaust
valve. h. Intake tube. i. Exhaust tube. j. Amplifier & headset.

Illustration of proposed insect ambush detector from the US Army. Credit: Clyde Barnhart, courtesy of the Aberdeen Proving Ground, Maryland.

after DDT's initial deluge, the scientists' interest dipped for about a decade, corresponding with the decimation of the bug. In the years that followed, what little research there was came from the developing world, where the pest was still a problem: countries in Africa and Asia, or institutions including the London School of Hygiene and Tropical Medicine, which often operated in those regions. Just a handful of studies were published each year. Many looked at the tropical bed bug rather than the common one, and most of the work focused on public health, pesticide effectiveness, or, as bed bugs evolved to withstand DDT and its cousins, pesticide resistance. The research on the latter grew in the decades following the war, for even DDT couldn't completely destroy what nature had perfected over millennia, and thus bed bugs hadn't disappeared entirely.

Four years after the Americans and the Brits added DDT to their wartime supply lists, scientists found bed bugs resistant to the insecticide in Pearl Harbor barracks. More resistant bed bugs soon showed up in Japan, Korea, Iran, Israel, French Guiana, and Columbus, Ohio. In 1958 James Busvine of the London School of

Hygiene and Tropical Medicine showed DDT resistance in bed bugs as well as cross-resistance to several similar pesticides, including a tenfold increase in resistance to a common organic one called pyrethrin. In 1964 scientists tested bed bugs that had proven resistant five years prior but had not been exposed to any insecticides since. The bugs still defied the DDT.

Soon there was a long list of other insect and arachnid species with an increasing immunity to DDT: lice, mosquitoes, house flies, fruit flies, cockroaches, ticks, and the tropical bed bug. In 1969 one entomology professor would write of the trend: "The events of the past 25 years have taught us that virtually any chemical control method we have devised for insects is eventually destined to become obsolete, and that insect control can never be static but must be in a dynamic state of constant evolution." In other words, in the race between chemical and insect, the insects always pull ahead.

• • •

Long before the army entomologists at Aberdeen were running the bed bug through a battery of detection tests, Robert Leslie Usinger, the Berkeley entomologist, had begun his work on his authoritative bed bug text, *Monograph of Cimicidae*. In the late forties, Usinger, a gentle and humble scientist by all accounts, proved to have an alter ego as a real-life entomological Indiana Jones. He inherited the cimicid monograph project from a curator of the Natural History Museum, London, who had to focus on other work. Cimicids were not Usinger's main entomological love—his favorite were the aquatic bugs whose larvae thrive in water—but he embraced the project with the same zeal as his many other books on insects, zoology, and natural history, and began to hunt the bed bug and its cousins across the globe.

The entomologists of this era traveled to the edges of the Earth just to collect a specific insect from a certain kind of place; the insects were, as one such explorer would later recall, the world's "smallest game." Usinger was no exception, honing his explorer skills from an early age, first as a lanky Boy Scout scanning the edges of the Grand Canyon for fossilized animal footprints and prehistoric Native American pueblos, and later as an undergraduate studying entomology at Berkeley.

Near the start of the Great Depression, in 1933, Usinger's mother dropped the budding entomologist and his college buddy at the edge of East Oakland, California, and watched nervously as they picked up their first ride as hitchhikers. The boys thumbed their way to Phoenix, Arizona; hopped a freight train to El Paso, Texas; crossed the river to Juárez; rode a train for three days to Mexico City; and then took a chicken-filled bus that would get six flat tires over the remaining eighty-four miles southeast to Real de Arriba, just outside of Temascaltepec. They did all this to spend a summer collecting insects for research. Since money was tight, Usinger funded the trip by peddling promises to Berkeley entomology students and professors that he would bring back Mexican insects for their respective research. This foresight earned the excursion $110. Around two months later, Usinger and his friend set out on a reverse journey back to California. When they arrived, they had one penny, around ten thousand insects, and one case of hepatitis (Usinger's, which set him back a year in his studies). Included in Usinger's private collection from the trip were some of his first bed bugs, which he had paid young boys in Tejupilco to gather for one centavo per batch of fifteen. He'd eventually use the specimens to study the structure of their chromosomes, which hold the DNA that make the bed bug a bed bug.

Fifteen years after his adventures in Mexico, now a thirty-six-year-old professor of entomology at Berkeley, Usinger prepared for what would perhaps be his most fantastic journey. It would comprise six expeditions and span five continents, over thousands of miles, and last nearly two decades—all to collect bed bugs, bat bugs, swallow bugs, and all other cimicids for his monograph. During his trip, he gained access to Europe's premiere institutions, poring over their bed bug collections. He visited bed bugs in the British Museum, as well as the Vienna Museum. In 1958 he even snuck from West Germany to East Germany to do research in the collections at the Berlin Museum. The city had been a center for bed bug research for many years, but access to the east had been cut off during the war. Traveling by car between East Berlin and West Berlin was still under tight control, but Usinger was able to take the subway for only a few cents and come up on the other side just blocks from the museum.

Usinger's field trips to trap live bugs took him to South America, where he stalked elusive bat bugs in hollowed Patagonian trees and waded in the nude through the water tunnels of Belém, Brazil, in search of more. In some cases, places that locals had said were teeming with bugs in recent years were barren, killed off by insecticide sprays. Still, he sought more and traveled throughout Africa including in the Belgian Congo, where his team crashed through dense forest with machetes in search of bat caves. He hunted the bugs on the other side of the world in Thailand and Japan, as well as at home in the United States. And in Egypt, he sweet-talked quarry workers into blasting the faces off ancient cliffs to access unreachable bat roosts and wandered through tunnels dug by grave robbers that led to the great pyramids of Giza, which held what he'd later call "one of the most remarkable bed bugs in the world."

Despite the occasional difficulty in finding the bugs, Usinger still found many, and from each site he airmailed live bugs back to his lab at Berkeley to set up colonies for experimentation. Included in these special deliveries were samples of the common bed bug, which his team of scientists let feed on bats, chickens, mice, pigeons, rabbits, and, when Usinger was available, a human. (He didn't feel right forcing the job on anyone else, so he hosted the insects himself.) The scientists sliced and diced some of the bugs in experiments; dried and flattened others on thick glass slides, their species and origin carefully recorded in pen; stabbed others with straight pins for easy display; and preserved still more in glass tubes filled with 70 percent alcohol, which they saved for later experiments. They tested insecticides, which clouded the lab with the pervasive smell of chemicals. They even played matchmaker, mating common bed bugs with pigeon bugs and other species to see if they could produce offspring and, if so, what those hybrids would look like. Years later I would find that to my untrained eyes, they just looked like more bed bugs.

All the while, as Usinger rifled through the museums, chased bugs in trees and caves, and toured with his family, who sometimes joined him, the entomologist carried a small vial of common bed bugs in the breast pocket of his shirt. He would periodically slip the vial out and feed the bugs on his arms and legs. This particular strain was part of an ongoing test to see whether initially feeding the bugs on one host and then switching to another for one or more generations would cause the younger bugs to prefer the newer host (they did not). Letting the bugs feed on anything or anyone but him would ruin the experiment.

But Usinger couldn't always cradle the precious vial safely in his pocket. Sleep, showers, and other daily necessities forced him to put it on a bedside table or dresser. During a family tour of Western Europe in the summer of 1959, Usinger left his globetrotting bed bugs in a hotel in Rothenburg, Germany. He didn't realize the mistake until the family had driven to Nuremberg, fifty miles to the east. He turned the car around. The family didn't arrive back at the hotel until late at night, and they found their room occupied by another guest. Fearing the hotel owner's reaction, Usinger decided not to report the missing vial and instead booked another room close by and waited until morning. As soon as the original room emptied, he posted Martha, his wife, as a guard in the hallway while he snuck back in. He couldn't see the vial anywhere. He fumbled around the room, retracing his steps, and finally found it wedged under the dresser. He slipped the bed bugs back into his pocket and the family resumed their tour. The same scenario played out in another hotel room in Cairo, where, while doing fieldwork on his beloved aquatic insects outside of the city, he realized the bed bugs were missing. Again he rushed back to rescue them and found the vial in the top dresser drawer in his room. And once in a London hotel, Usinger left the vial on his dresser when he went out for the day. When he returned, a familiar smell hit his nostrils. Fly spray. Most likely laced with DDT. His bed bugs were only spared because their container was so tightly sealed; he later told the hotel management that he was allergic to fly spray, a white lie.

Around seventeen years after he set out on his first *Monograph* trip, the fifty-two-year-old Usinger hid away in a university lab outside of Palm Springs. There in the California sands with "nothing but cactus and desert all around"—alone with his books, journals, and extensive notes—he worked day and night, finishing his manuscript in just one month. Half a year later, although his monograph hadn't even made it through the printers, he was already plotting a fresh bed bug excursion to Moravia, a stretch of land in Eastern Europe in what was then Czechoslovakia.

According to a Czech entomologist named Dalibor Povolný, whom Usinger had befriended and collaborated with over the years, the region formed a natural line separating the distribution of the common bed bug to the north and a bat bug species to the south. This invisible ecological divide intrigued Usinger. Why did it bisect the living spaces of these cimicids just so? Povolný's work, too, showed the common bed bug living in bat caves in Afghanistan and in churches throughout Czechoslovakia. Usinger took this as evidence of the bed bug's origin in bat caves, and he was anxious to see these examples in person. Usinger continued plans for the trip over the next two years, sorting out his passport and making additional arrangements to attend an entomological meeting in Moscow on the invitation from the Academy of Sciences of the USSR, as well as a side trip to collect insects in Poland. As he planned his trip, he fell gravely ill. Doctors would eventually diagnose him with cancer. Decades later his son would tell me that the plan for the adventure in Eastern Europe and the USSR was likely a mere distraction from two failed surgeries and cobalt therapy.

• • •

When I purchased my own copy of Usinger's plain, heavy book more than four decades later, I skipped the preface to thumb through the meatier sections on the bed bug's history, ecology, and biology. I spent time to marvel over the cimicid portraits. I read and reread the history of the bed bug, amazed at the way it radiated out from the Mediterranean, shadowing our every move. It was only months later that I turned to the earlier pages on a whim and discovered the two most captivating sentences out of tens of thousands of words. These described Usinger's research in Patagonia and the pyramids of Egypt, and mentioned the help he'd received from wood cutters who felled huge mangrove trees filled with fish-eating bats in Trinidad and mountaineers who roped "down vertical cliffs to the nests of white-throated swifts in California."

I next did something that Usinger himself would have likely never dreamed possible: I typed "Robert Leslie Usinger entomologist" on my computer, which spat out his abbreviated life history as well as photos of a fair-haired man with a crooked smile in a series of fifties-style suits and skinny ties. In one, he posed with a microscope and looked to his left at someone off-camera, a pale eyebrow slightly raised. A formal portrait showed him smiling a toothy grin directly at the camera, his Ivy League haircut freshly coiffed. And a full-length shot in front of a tall hedge showed him casually clasping his hands at his waist and squinting in the sun. I tried to picture this clean-cut scientist hacking through forests and wriggling through caves, but I could not. As I scrolled through his pages on the Berkeley website, I learned that the school still held thousands of his bed bugs at its Essig Museum of Entomology—just a fraction of the 61,919 true bugs that comprised his full collection—and that the museum was open to the public by appointment. I was already traveling to California the following month. I booked a day at the museum.

After wandering Berkeley's sprawling grounds in search of the Valley Life Sciences Building and getting lost on paths winding through groves of eucalyptus and oak, I found the door to the Essig Museum at the end of a dark and sterile hallway. I opened it and stepped into a long room that smelled faintly of rubbing alcohol. A few researchers were hunched over microscopes in a small alcove to my left. A man jumped up when I said hello, shook my hand, and guided me to a guest book to sign in. As I scrawled my name, I saw that the last entry was from a week prior. Next, he led me across the room, past dozens of stacks of wooden trays with glass lids from which I glimpsed the familiar shiny shells of *Coleoptera* (beetles), the delicately painted *Lepidoptera* (butterflies and moths), the cellophane wings of *Odonata* (damselflies and dragonflies), and hundreds of other insects frozen in place and time with simple straight pins.

"Do you need a microscope?" the researcher asked. I told him I did not, that I was there to see the bed bugs, but not exactly to study them. I wasn't doing science. He peered at me over his glasses, eyebrows raised. An unspoken question hung in the air: *Why, then, was I there?*

"I'm a writer," I started. And stopped. It was early in my research, before I had a publisher, before my book proposal was much more than a few rambling pages. My spiel was unpracticed. "I want to write a book about bed bugs and—well—I'm doing research for it, and Usinger's *Monograph* led me here and I was in the neighborhood so I thought . . ." A jittery explanation, but he

Robert Usinger. Credit: Essig Museum of Entomology, University of California, Berkeley.

nodded thoughtfully and told me a story about an entomologist he knew who, while stationed on a South Pacific island, had dragged his bed away from the wall and arranged all the sheets so his bedroom's many bed bugs couldn't climb in and get him. Instead, the bugs crawled up the walls to the ceiling and dropped on him from above. *Everyone has a bed bug story*, I thought, as the researcher pointed me down one of the neat corridors of stacked insect trays, toward a bookshelf. "Usinger's bed bugs are there."

They weren't in wooden trays like the *Coleoptera*, *Lepidoptera*, and *Odonata*. Instead, he had pointed me to a set of fifty-three black faux-leather books. I walked down the corridor, pulled one off the shelf, and unclasped its two gold fasteners, unsure of what I would find inside. Dead bed bugs pinned to downy white cotton? Floating in small vials of alcohol? Dried and loose, I wondered wildly, poised to flutter to the floor and cover my shoes? I swung it open. It wasn't a book. It was a box, with a hundred numbered wooden slots filled with heavy glass slides. At the center of each was a generous dollop of yellowed resin capturing a bed bug or bat bug or pigeon bug or their cousins, flattened as though they had been run over with a miniature steamroller, so thin that the bodies were, in places, transparent. Each was a perfect template for a detailed drawing, and I wondered if it was possible to match them to the portraits in Usinger's book. Handwritten labels—some with entire sentences or paragraphs telling fragmented stories—were attached with clear tape.

One read:

Western desert 4 kilom
West of El Mansuriya
Giza, Egypt
H. Hoogstraal

And another:

Grotte salo premi at Kako,
a suburb of Jadoville, Katanya,
in crevices on ceilings of damp walls,
complete darkness, 33°C
R. Leach, R. Fareaux
1-6-58
Bats Present.

And another:

C. lectularius
Host: Indian Student
Chiapas, Mexico

I sank to the floor, lost in the descriptions. There were entire boxes labeled *Human colony* and *Bat to human*; geographical descriptions as far-flung as Argentina, Brazil, Czechoslovakia, the Congo, Egypt, Finland, Germany, Japan, Java, Kenya, Minnesota, Mongolia, Napa Valley, Nevada, Russia, Sierra Leone, South Africa, Thailand, Vietnam, Yemen; collections of the common bed bug and pigeon bug cross-breeding experiments, with both parents sharing a slide and the succeeding slides holding generations of their offspring, a series of twisted family portraits. There were slides with an unnamed species from the genus *Primicimex* from

Bed bug slides from the Robert Usinger's collection. Credit: Essig Museum of Entomology, University of California, Berkeley.

Dalcahue, Chile, a bed bug so big that when I placed a penny next to it, the comparison revealed that its body alone could have obscured Abraham Lincoln's head. Some of the labels clearly came from other scientists, but many had Usinger's name on them. His handwriting was tiny and delicate, the paper labels browned with age. It struck me that he had written those words with his own hand, and reading them was like going back in time.

• • •

Just seven years after Usinger published his monograph, another young scientist on the other side of the country would similarly chance upon an unusual bed bug research project—one that would prove instrumental in unraveling how the bug developed pesticide resistance, among other things. In the early seventies, Harold

Harlan was just starting out as an entomologist with the US Army, a job that he would have for twenty-five years and that would eventually take him on tours of duty to Vietnam, Panama, and Saudi Arabia. In 1973 he was stationed with a preventative medicine group at a less exotic locale: Fort Dix in central western New Jersey.

One cool spring day, recruits who were stationed at the fort for basic training woke up with insect bites. As the resident entomologist, it was Harlan's job to sleuth out the cause. He inspected the rows of red welts on the recruits' arms, legs, and torsos. The patterns and locations didn't suggest mosquitoes or any other biting insect common to New Jersey. The frosty March nights were too cold for mosquitoes, anyway. The recruits thought they were attacked in the early morning hours as they slept, since they didn't notice anything biting them during the day as they sweated their way through training courses, and thus Harlan's detective work eventually led to the young men's sleeping quarters in lightly painted cinder-block barracks. He searched in and around the beds and found nothing. Next, he peered into the fine cracks and pores in one of the concrete walls near a bunk. Wide-eyed flat bugs peered back at him. It was the first time he saw bed bugs that weren't mounted on a slide or sketched in a textbook, true for most students and young entomologists at the time. He checked the walls in the other bedrooms on the first floor of the barracks and found infestations in five of eight.

Because bed bugs were so rare, the army tasked Harlan with writing a scientific note describing the encounter, which was published the following year in the journal *Military Medicine*. He was also responsible for arranging for the elimination of the bed bugs. He reported the infestation to the Army's Pest Control Section, run by civilian exterminators. They doused the beds and wall cracks of the Fort Dix barracks with diazinon, an organophosphate later banned for indoor use in 2004 along with a slew of related chemicals because of health and environmental concerns. (The compounds were also generally disliked because of their pungent, garlicky smell.) Before the pest control operators came in with the poison, Harlan was saddened by the bugs' impending death. He was drawn by their cryptic behavior and their novelty. They were rare game—the black rhinos of the pest insect world—and he wanted to keep and study them. And so he quietly spared around two hundred of the bugs by herding them into a Mason jar and taking them home.

To study these rare bed bugs, Harlan had to keep them alive. These were *C. lectularius*, the common bed bug, and their livelihood depended on blood. Harlan figured his own was as good as any, and so he stretched a pair of his wife's old nylons taut across the mouth of the jar and then held it against his arms and legs, allowing the bugs to suck his blood while preventing their escape. The panty hose tore too easily for comfort—Harlan liked the bugs, but not enough to set them loose in his house, although that did accidentally happen three times—and so he switched to Dacron netting, made from a strong synthetic fiber that is also used to make boat sails and kite strings, and, later, to a fine copper wire mesh. Harlan placed crumpled sheets of white lab filter paper inside the jar, which gave the bugs something to crawl on other than slippery glass.

Over time his bed bug homes evolved until he was spending hours laboriously designing shelters made of thin cardboard, cutting deep ridges along the opposite edges and then folding the pieces into tight accordions to increase the surface area. The bed bugs could climb an endless loop over the peaks and into the valleys of these dainty mountainscapes. They could also flatten their bodies into the folds. Years went by, and then decades, Harlan's collection ballooning at its peak to twenty-five thousand bed bugs spread among thirty or so pint- and quart-size jars in the modest Maryland home that he shared with his wife and two sons. His colleagues thought he was nuts. Neither they nor he knew it at the time, but those bugs and their offspring would eventually prove vital to entomologists and exterminators across the world.

• • •

Meanwhile, a pesticide revolution was under way. It started as a seed of doubt among entomologists and ecologists in the forties and fifties as they debated the merits and drawbacks of DDT in the pages of dry scientific journals. The experts were particularly concerned about the compound's persistence in the environment from its residual staying power, the very thing that made it so helpful for killing bed bugs and a feature that even Paul Hermann

Müller had fretted over two decades earlier. They were also worried about the pesticide's unintended threat to insects and larger
animals that were not its target, including humans.

It wasn't until 1962, though, that the revolution bloomed. That
summer, a young biologist and science writer named Rachel
Carson pushed the debate into public consciousness through a
three-part series in the *New Yorker*. In September she published
a more complete version in a book called *Silent Spring*. Soon the
book would spend thirty-one weeks on the *New York Times* best-
sellers list, galvanize the fledgling environmental movement, and
become a topic of great interest among both scientists and the
public. Even in the Usinger household, the entomologist and his
friends and colleagues discussed Carson's work during parties.

Carson's point was simple. By dumping tons of pesticides indiscriminately on orchards and beyond, humans were inadvertently
changing the world's ecology. DDT and the more than two hundred other insecticides, herbicides, and rodenticides developed
in the period between World War II and *Silent Spring*'s publication
traveled our water- and airways. Winds carried DDT from orchards
to nearby ranches, forests, and neighborhoods. Rains soaked it
into the soil, where it entered the groundwater, and waterways
washed it downstream. Its ability to travel long distances in the
upper atmosphere eventually spread it even as far as the Earth's
frozen poles. In its travels, DDT poisoned non-pest insects, including the pests' natural predators. And many of those unintentionally affected insects were food for larger, more complex animals
including frogs, fish, and birds. DDT lingered in the fatty tissues
of these creatures, its concentration increasing as they ate more
and more pesticide-laced insects. As the chemical accumulated
in the animals' bodies, wrote Carson, it altered their nervous systems, reproductive systems, and hastened their death.

DDT was in human bodies, too. The higher up on the food chain,
the higher the concentration; after consuming fish that had eaten
DDT-afflicted bugs, fruit coated in DDT residue, and beef or milk
from cattle that had been chewing grasses grown in DDT-soaked
soil, people were threatened with the highest concentrations of all.
Indeed, traces of the chemical persisted in human bodies for a long
time, stored in fat tissue and breast milk. But whether DDT's pres-

Rachel Carson. Credit: Joanna Barnum, joannabarnum.com.

ence posed a definite serious threat isn't clear. Immediate effects from an overdose were known: while the acute toxicity for mammals, including humans, was relatively low, direct contact with DDT could spark tremors, joint pain, nervousness, and depression. The long-term impact from indirect ingestion was less conclusive, in part because the military researchers who approved it to wipe out wartime diseases hadn't run tests on chronic exposure. In animal studies, DDT increased the likelihood of liver tumors; exposure in humans has been linked to higher rates of breast cancer, although it's never been proven as a cause. The pesticide is also a possible endocrine disruptor, which means it might interfere with hormone signals that are key in sexual development and reproductive health. Today, based on this scant evidence, the official stance of the US government and many other nations is that DDT is probably, but not definitely, carcinogenic.

Despite these hints of biological danger, even Carson argued that pesticides are sometimes necessary, writing in the opening pages of her book: "It is not my contention that chemical insecticides must never be used. I do contend that we have put poisonous and biologically potent chemicals indiscriminately into the hands of persons largely or wholly ignorant of their potentials for harm." Carson advocated not for the end of DDT, but for a more targeted, thoughtful approach; it wasn't the pesticide itself that posed an ecological threat, but the way people were overusing it. As the physician Paracelsus stated during the Renaissance, and toxicologists today repeat like a mantra, it's the dose that makes the poison.

In the decade that followed the publication of *Silent Spring*, debates on DDT's fate raged within and between scientific communities, the chemical industry, and government regulatory agencies. The USDA continued restrictions of DDT through the sixties, a process it had started the previous decade. By 1970 the newly formed Environmental Protection Agency took over these responsibilities from the USDA. Two years later, the government decided to change the toothless Federal Insecticide, Fungicide, and Rodenticide Act of 1947, which had been written with the intention of forcing chemical companies to prove that their insecticides were safe but hadn't really worked.

By 1972, after lengthy and vitriolic public hearings, DDT was banned for use in the United States. But it wasn't illegal for the chemical companies to manufacture or export the pesticide, and for some, the DDT business boomed. Montrose Chemical Corporation of California, the last American DDT producer, had its most profitable years directly following the 1972 decision. Montrose didn't stop production until 1982; the following year, it shuttered its last factory on the outskirts of Los Angeles. Today that factory is a Superfund site, part of the US government's program to clean up uncontrolled hazardous waste.

With DDT still in use throughout most of the world in the seventies, eighties, and nineties, and with other synthetic chlorinated hydrocarbon pesticides still available in the United States and beyond, insect resistance continued to build. Bed bugs were no exception. People used these pesticides for bed bugs in regions outside of the United States where the pest was still common, and also inadvertently dosed the bugs while treating for other insects. Bed bug insecticide resistance grew, for example, in malaria-ridden parts of Africa and Central America as the World Health Organization tried to curb mosquitoes by treating homes with DDT. All it would take for the bed bug to roar back would be a way for it to spread from those resistant hotspots to the rest of the world.

FOUR

THE RETURN
A Pyrethroid Paradox

The Chernuchin Theatre on Fifty-Fourth Street in Manhattan had bed bugs. They wore tight shiny costumes made out of fabric rescued from a dumpster located between Juilliard and the New York City Ballet, and wild helmets with antennae, buggy eyes, and sharp teeth cobbled together with recyclables and dollar-store purchases. They sang and danced in a circle around a disheveled woman in a lab coat who was chained to a stack of dirty mattresses.

The most magnificent bed bug of all, whose name was Cimex according to the Xeroxed program spread in my lap, darkly serenaded his hostage as she strained against her bonds. His costume would have made a *Labyrinth*-era David Bowie very proud. The actor playing Cimex towered over six feet and wore bronze novelty spandex trousers with a sequined codpiece and a metallic fringed mesh top. A cape made of a recycled belly dancer's skirt and more than two dozen black trash bags cascaded from a bare set of football shoulder pads, which were spray-painted bronze and spiked with two-dozen bejeweled strands of music wire that shivered with each hip swivel. His headdress anchored foot-long antennae that reached to the sky and copper horsehair dreadlocks that hung down his back. He strutted groin-first as though led by his sequined cup as it twinkled in the stage lights, heavy boots clomping, dreadlocks and cape swinging, lyrically begging the restrained scientist to be his bed bug queen.

Of course, the Chernuchin didn't actually have bed bugs, although the duct-taped upholstery in a front-row seat looked suspect. Instead, the theater had *BEDBUGS!!!*, the eponymous Off-Off-Broadway sci-fi rock opera running a two-week showcase during

BEDBUGS!!! The Comedy Sci-fi Thriller Rock Musical. Credit: Rex Bonomelli.

Hurricane Sandy. Because the subway was still shut down in the storm's aftermath, I had ridden my bike from Brooklyn to the theater district, one hour each way on a cold November Saturday afternoon. As I puffed over the Brooklyn Bridge, I wondered if I wasn't taking this whole bed bug obsession a little too far.

The musical did not disappoint. I sat grinning throughout it, doubts cast aside, scribbling notes in a small pad on my lap. In addition to the giant bed bugs and the imprisoned scientist, who had inadvertently forced their mutation into flesh-eating giants with an aggressive pesticide made from pyrethroids and rotten coconut, it featured a musical number titled "Silent Spring" sung by an actor dressed as Rachel Carson's author photo; a sprinkling of scientific-ish facts about bed bug sex and pesticides; an archetypal mad scientist in a long white coat tossing colorful poisons together in her lab; and a Celine Dion–inspired character in drag. With some luck and roughly a half million dollars, the show's creators aimed to bring it to an open-ended Off-Broadway run by the end of 2014. After that: Hollywood.

The musical wasn't the first modern ode to the bed bug. In 2011 in Manhattan, a group of Broadway pit musicians performed the incidental music to a famous Russian play called *The Bedbug*, a futurist satire about socialist utopia, love, and a time-traveling bed bug, published in 1929 by the poet Vladimir Mayakovsky. The incidental music was written by the Russian composer Dmitri Shostakovich in the same year. In 2012 I saw Brooklyn hipsters stage the play in a small theater in Bushwick, a neighborhood with bona fide bed bugs raging through its communal artist lofts and housing projects.

Around the same time that I saw the play, I discovered a bed bug music revival on the Internet beyond what recording artists in the twenties and thirties could have imagined. Amateur and professional singers were posting user-generated content on YouTube and MySpace, including multiple covers of Bessie Smith's "Mean Old Bed Bug Blues" and new music collectively representing hip-hop, reggae, pop, punk rock, rhythm and blues, country, German synthpop, swing, jam bands, and indie rock—so many songs that I collected the best on a blog that I was able to update with new material every week for more than a year.

There was bed bug art, too, including a series of graffiti pieces by the New York–based artist Samuel Mark, which he drew on alley walls and discarded couches and mattresses across the city. Bed bugs also appeared in comics and graphic novels, which I read with glee, including the tongue-in-cheek Florida zine *Save the Bedbugs* and the cockroach-themed series *The Exterminators* published

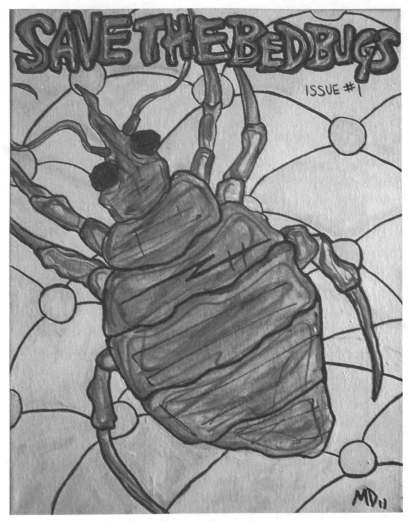

Save the Bedbugs, Issue #1 cover. Credit: Matt Deterior.

by Vertigo, a DC Comics imprint. Inexplicably, there was also an installment of a graphic novel series in which the Three Stooges are bed-bug-fighting pest control operators. I found low-budget bed bug movies and comedy pieces, such as an Isabella Rossellini *Green Porno* episode in which the actress performs an artistic interpretation of bed bug sex. I also saw the insects featured in popular sitcoms such as *King of Queens, 30 Rock,* and *The Office,* the latter in an episode where Dwight Schrute describes them as smug, self-assured pests who think "everything's a joke." And in 2011 a supernatural thriller novel—with the classic plot: young family moves into too-good-to-be-true home and horror ensues—chronicled a possibly demonic bed bug infestation in a Brooklyn brownstone. Within a year, Warner Brothers optioned the story for a film adaptation.

As I marveled at the modern bed bug renaissance, I wondered: *How did this happen? How did we go from having so few bed bugs that we forgot they existed to the nearly sold-out musical BEDBUGS!!!, revivals of an eighty-five-year-old Russian play, and a horror flick picked up by a major movie studio?*

• • •

The explosion of bed bug art and literature followed a surge in media coverage; the bug's foreignness and the ease of its spread provided fodder for the playful and dark bed bug catharsis, glorious bed bug rock opera and all. Media stories first appeared in the late nineties and early aughts in newspapers from the major cities getting hit by the resurgence: the *Guardian* in London, the *Daily Telegraph* in Sydney, and the *New York Times*. Most were stories about the bed bugs showing up in both budget and fancy hotels. The *Times* began consistent bed bug coverage in 2005, moving from articles on the new plague to increasingly obscure items about bed bugs attacking public libraries, the first-ever national American bed bug summit in Chicago, and the newspaper's trademark trend pieces, including one on the loneliness and shame of an infestation and its cloud over personal relationships.

As the bed bugs spread, smaller newspapers picked up on the trend in Atlanta, Austin, Santa Fe, San Francisco, Cincinnati, and throughout Florida. Scanning these news archives on my computer, I was amazed at how many stories had been published both before and contemporaneously to the 2005 *New Yorker* piece about bed-bug-ridden college girls that had given me comfort months after my first bed bug encounter in the city.

In 2010, one year after I suffered through my second and third brushes with bed bugs during a series of Brooklyn summer heat waves, the mainstream media storm hit its peak, breathlessly re-

"The Tenting," a comic from *Save the Bedbugs*. Credit: Mike Seyler.

porting infestations across New York in the Empire State Building, the United Nations, a Victoria's Secret store, the Waldorf Astoria, and the New York Public Library. CBS called 2010 the "Year of the Bed Bug" and named the pest among the biggest health stories of the year. Although these weren't the first media reports, for many people the tales were an introduction to the bugs. The fear of an alien pest infiltrating the sanctuary of the bed passed quickly from person to person thanks to the Internet, where stories both true

and false are shared with a simple click of a mouse or swipe of a screen.

The actual bed bug resurgence likely began several years before even the earliest media stories. Isolated cases popped up in the mid-eighties and nineties along the densely populated Northeast coast and in hub cities in the United States, with similar patterns in the other resurgence zones worldwide. One early documented case I found in New York is from 1996 and comes from a hostel on

Bed bug graffiti in New York City. Credit: Samuel Mark.

the Upper West Side. There, a Manhattan exterminator named Jeff Eisenberg tried to uncover the source of mysterious bites, which he initially thought might be fleas, ticks, or mosquitoes. Or, when those didn't turn up, he wondered if it was pigeon mites. While weaving through floors packed with international travelers, rows of bunk beds, well-loved sleeping bags, and oversize trekking backpacks, he discovered secretive bugs hidden deep in the bed joints and in the folds of the sheets. An entomologist confirmed they were bed bugs but couldn't advise on how to kill them, and when Eisenberg called retired colleagues who were old enough to have treated for bed bugs decades earlier, they gave useless advice on pesticides that were no longer available or legal. For months, the hostel's bugs remained despite all efforts, possibly due to the constant turnover of global travelers.

Exterminators were similarly called to treat a mysterious bed bug infestation in a summer camp tucked in the mountains of western Maine in 1997 and an outbreak in a popular commuter hotel in Secaucus, New Jersey, in 1999. Also in 1999, the US technical director for Orkin, a mustachioed man named Frank Meek, found bed bugs in his hotel room during a trip to Orlando, Florida. He called his company's local branch to report the problem and get the bugs treated. "How do we do that?" they asked. "Beats the hell out of me," he replied.

Meek at least knew what he was dealing with. Most people were not so lucky, which prolonged their diagnosis. The crawling things they found hiding in their beds, they decided, were ticks or baby roaches or, perhaps, some sort of mite. Or they reasoned that they'd stayed out too long at dusk when the mosquitoes were hungry. Or they claimed it was spiders, which are perhaps the most frequently accused arthropods in the animal kingdom. Spiders may be falsely blamed for medical maladies from small insect bites and poison ivy rashes to bacterial and fungal infections to ulcers, herpes, and skin cancer. I'm guilty of this. Not only did I pin my first bed bug case on ravenous imaginary spiders hiding in my apartment walls, but I also thought I'd been bitten by a similar strain after an afternoon nap in a Spanish hostel in 2002, when I awoke to a straight line of throbbing bites on my arm (which in retrospect may have been the work of bed bugs). The reality is that while some spiders do bite, it isn't that common and it is usually an act of self-defense. Out of the 44,000 known spider species, likely just a quarter of the actual number on this planet, not one feeds on human blood.

When faced with the harsh truth that I had bed bugs and wasn't suffering from a particularly bad mosquito year or preternatural bloodthirsty spiders, I denied it. Others did, too. *Bed bugs! Prepos-*

terous! They're not real! In some cases, people ignored the evidence for weeks or months or years and subsequently gifted infestations to family, friends, neighbors, and possibly anyone who slept in a hotel room after them. Finally, some called an exterminator. Before long, pest control workers in those original coastal and urban hot zones connected the red itchy dots to recognize that bed bugs were spreading like a pox across their cities. But the bugs had been gone so long that only the old-timers had ever heard of the pest let alone killed it, and none had passed the skill to the younger generation inheriting family businesses or starting up new ones.

The pest control industry had ideas on how they had left our homes open to the unprecedented bed bug attacks. By the nineties or so, exterminating tactics had shifted from regular baseboard sprays in public housing and apartment buildings or occasional treatments in single-family homes. The new approach was bait, a poison especially common for indoor urban pests like the cockroach. Roaches eat, among other things, our dropped crumbs and garbage; if lured by a poisoned morsel, they will eat that, too. They will then skitter back to their nest, curl up, and die, possibly first squeezing out poison-laced feces. Nearby roaches nibble the dead body and tainted poop until they also drop dead. And so on. But blood-feeding insects are not so readily seduced by fake food. Their meals flow inside living beings and aren't easy to poison.

According to the bait hypothesis, the shift from home pesticide sprays created a safe haven in which bed bugs thrived. This idea assumes that the sprays previously used for cockroaches might have curbed bed bugs, which, depending on where they were sprayed, may not have helped. Cockroaches like the crumby counters and torn food packages of the kitchen, or the dank corners or shallow puddles of water left from freshly showered feet in the bathroom, which is where an exterminator sprays for them. Bed bugs prefer to be near people at rest, which, unless it's a particularly rough night, isn't the kitchen or bathroom floor.

Regardless of whether baits were to blame, exterminators needed to move back to sprays in order to treat the surging bed bugs. When the treatments became more common in the early aughts, there weren't many options. The United States had just one modern pesticide class, pyrethroids, that could be legally sprayed in the bedroom; abroad, pyrethroids were one of just a few choices. Pyrethroids had risen in popularity after DDT's demise, and they were closely related to pyrethrins, the ancient compounds found in chrysanthemum flowers. The flowers were first crushed into powder and used as an insecticide in what is now Iran around 1800, and by the nineteenth century, it was used throughout Europe. The powder quickly paralyzed bed bugs but was relatively harmless to humans and household pets, which helped its popularity. The drawback was its instability in sunlight, which meant it required vigilant reapplication.

In the modern version, the suffix -oid means "resembling" or "like," which comes to the English language from Greek. Pyrethroids are simply a synthetic version of pyrethrins; while both have virtually identical chemical structures and functions, the former is constructed in a lab rather than inside of a flower. Synthetic chemicals allow for extra control, for an engineer can tweak the details of the chemical's shape and thus its function. Pyrethroids are designed to be more stable in light to have the residual power necessary for bed bug control. Both pyrethrins and pyrethroids kill by switching the insect's nerve signals permanently on until they fire hundreds or thousands of times too many, frying the bug's nervous system and causing tremors, convulsions, paralysis, and death. If this sounds familiar, it's because the chemicals appear to work along the same molecular pathway as DDT.

Americans weren't dousing pyrethroids and their natural counterpart, which was also still in use, with the same unbridled zeal seen with DDT. Still, the pesticides were popular and remain so. As of 2013, there were more than 3,500 products containing pyrethroids and pyrethrins on the US market alone, including over-the-counter insecticides such as Black Flag and Raid, professional-grade formulations, dust mite mattress covers, flea and tick pet treatments, scabies and lice creams, and agricultural and garden pesticides. I remember watching as an exterminator doused my mattress with the stuff and wanting him to dump on more, more, more, anything to get rid of my tormentors. Across the country, pest controllers treating for a bed bug infestation, lacking other options, would also grab a pyrethroid, spray the hell out of the bedroom and the bed within it, and hope for the best.

But the pyrethroids weren't always working.

• • •

As the pest controllers battled their first bed bugs, urban entomologists didn't yet know about the infestations, let alone the pyrethroid problem. Most were making a living studying more common urban banes such as ants, roaches, or termites, and few had thought much about bed bugs since Robert Usinger's work, if they had even been alive and practicing that long ago. Just as in the pest control industry, the virtual absence of the bugs meant little incentive for research.

That changed around the early 2000s when pest control operators and industry people began talking to academics about what they saw as an alarming increase in bed bug cases. One was Rick Cooper, the vice president and technical director of Cooper Pest, a second-generation pest control company that his father had opened in Lawrenceville, New Jersey, in 1955. A quiet man with unruly dark hair and glasses, Rick, along with his older brother, grew up working for the family business during summer breaks from school, and the brothers took it over in 1991.

Rick was relatively unusual amongst his exterminating peers, with a keen interest in insects that eventually led him to a doctoral program in entomology at Rutgers University. But despite growing up fighting urban pests and spending years in the lab, Rick never saw a bed bug apart from those preserved in alcohol in taxonomy class. That changed in 1999 when he treated an infested hotel. Within a year, he had half a dozen more bed bug cases; within two years, the calls were steady; within three or four, he was presenting on bed bugs at US industry conferences, where most of his fellow pest controllers had yet to see a bed bug because the problem was, so far, mainly geographically isolated to the East Coast.

It's not unusual for academics to attend these meetings, and one from the University of Kentucky, an entomologist named Mike Potter, noticed Cooper's presentation and pulled him aside in the hallway between sessions. The university entomologist acknowledged that this bed bug thing was interesting, but said he just wasn't experiencing the same problem in his part of the country. The two parted ways. Months later Potter called Cooper with a fresh perspective: "Okay, Rick," he recalls saying, "tell me everything you can about bed bugs. Haven't had a call on them for nineteen years, but now the phone started ringing a month ago and it hasn't stopped." Bed bugs had hit Lexington.

By 2004 a good friend convinced Cooper to present his work to a different sort of audience: the biennial National Conference on Urban Entomology in Phoenix, Arizona. There in a Hyatt conference room shaded from the hot desert air, Cooper shifted nervously in front of around seventy urban entomologists before launching into an impassioned plea for someone—anyone—in the audience to start studying bed bugs.

Cooper flipped through PowerPoint slides of rough numbers on the bugs' spread through the Northeast, Florida, and California, and argued they would soon pour beyond those regions and fill in the rest of the country. His fifteen minutes ran out before he could open the floor to questions, but soon he was surrounded by scientists peppering him with *wheres* and *whys* and *hows*. He had no answers. Most of his audience left interested but skeptical. "Bed bugs won't come back," they said. "Neat story, but not worthy of research dollars," most seemed to agree. "You're fearmongering" was the subtext. But a few others thought Cooper was right. One was the same Orkin director who had found bed bugs in a Florida hotel years earlier, who also had given a talk about the bed bug's comeback at the meeting in Phoenix. Another was an entomologist named Dini Miller, who began setting up her own bed bug lab at Virginia Tech later that year. And Potter and his colleagues at the University of Kentucky, in particular an expert in insect behavior and chemical ecology named Kenneth Haynes, also started planning their bed bug lab. The scientists from Kentucky and Virginia would soon have the first formal, active bed bug research labs in the country in decades.

The scientists sought to answer: *Where had these bed bugs come from? How were they able to spread so quickly? And why were they so difficult to kill with pyrethroids?* To start, the researchers had to dig back into old literature to learn about the bugs and try to pick up where it left off. This led Potter, Miller, and others to papers published in the thirties and forties, as well as to Usinger's *Monograph*, the most recent significant work, which was out of print. Some of the scientists were able to borrow worn copies through inter-library loans, but others were faced with paying up to $300 for the 1966 edition. As the scientists studied the texts, a slew of conjectures on the origins of the new resistant bed bug emerged.

One thing was clear: bed bugs follow humans. We are their food, and thus they have evolved to accompany us as we move about the world. This pattern of travel is one the insects have performed for at least thousands of years, as Usinger and others have pointed out. Even after the widespread use of DDT, some bed bugs survived and they found pockets of humanity on which to feed. But not everyone agreed about where those reservoirs existed and how the bugs spilled beyond them.

In the United States, one hypothesis suggested, the bugs were homegrown, perhaps surviving in communities overlooked by most of society, such as subsidized housing and homeless shelters. Anecdotal evidence supports this idea, with reports and photos of bed bugs ravaging people in projects from Harlem to Washington, DC, in the decades following DDT's rise. Without money and community support, even this miracle pesticide was ineffective. Or perhaps the bed bugs found refuge near non-human hosts, surviving on chicken blood on poultry farms scattered throughout the country. In either case, all the bugs needed to break out of these reservoirs was to hitchhike, moving from one community to another on the cuff of a pant leg or in the crease of a backpack of a person visiting family or friends, or going to work, or moving up the social strata to middle-class apartments or homes. Or maybe the bugs moved from chicken house to farmhouse to pickup truck and beyond. As the people from the bed bug reservoirs went about their daily routines, the bugs would have spread, perhaps abetted by the changes from pesticide sprays to bait.

There are more efficient vehicles for dispersing bed bugs, other experts countered, as well as possible reservoirs situated beyond geographical confines. One such vehicle: the airplane. One such reservoir: the entire rest of the world. In 1978 the US Congress passed the Airline Deregulation Act, which removed the government's control over key aspects of the industry: pricing, routes, and schedules. This allowed new airlines to come to market and also encouraged existing companies to add new planes to their fleets. By the early eighties, as the laws phased in, the airlines quickly learned that people liked cheap tickets. To lure customers, competitors undercut one another. Cheaper tickets and more available routes helped boost domestic travel; in the decade following the act, annual flights within the United States more than doubled despite the fact that high fuel prices hurt the industry overall.

In the early nineties, partially influenced by the momentum of American airline deregulation, a similar phenomenon happened for international flights through the first open skies agreements— policies that pushed global commercial flights toward a free market system. Since then, the United States has inked these agreements with more than 130 nations. Between the early nineties and early aughts, flights to and from the United States from virtually all parts of the world increased, as did traffic between dozens of other countries. With more airplanes crisscrossing the globe carrying more people from more places, each taking suitcases and trunks and carry-ons stuffed with clothing, bedding, and gifts to and from their home cities and villages, there were more opportunities for bed bugs to hitchhike.

Of course, these are just a few possible strands of the bed bug resurgence narrative. An increase in global mingling, other experts claimed, could also be attributed to modern-day wars, with soldiers stationed around the world for extended periods of time and particularly in the Middle East, the bed bug's alleged region of origin. Others pointed out the increase in emigration and subsequent travel back to motherlands to visit family and bring gifts, with return trips laden with comforting, familiar items to take away the sting of homesickness.

Also plausible: the role of growing worldwide wealth and a new global economy. People with more money can buy more things, which are increasingly manufactured in factories scattered all over the world and then shipped by plane or train or boat to accumulate in closets and dressers and bedside table drawers. Those companies send managers to check on overseas factories and processing centers, requiring plane rides and hotel rooms. On a more local scale, an increase in manufacturing and purchasing may have led to more people replacing old furniture, which they leave on the sidewalk for the garbage truck or someone looking for a free couch.

On a broader scale, more and more people are moving to cosmopolitan areas worldwide, cramming closer and closer together and making it easier for bed bugs to transmit from one home to the next. More than half of the planet's 7 billion people live in cities. There are also simply more of us than there used to be, as the world population has nearly tripled since 1950. Not only have we provided better opportunities for bed bugs to spread, whether by global trade or consumerism or city living; we've also increased their food supply.

• • •

Before the scientists could test these hypotheses to see where the bed bug resurgence began, or try to understand the bug's biology to find new ways to kill it, they needed to set up their labs. Examining the bed bug would eventually require thousands to hundreds of thousands of insects to poke and prod and poison—enough for hundreds per experiment, with the ability to quickly replenish the stock. The scientists had to learn to hand-raise healthy bed bugs that would happily breed to make more, which meant giving the insects what they seek in their natural milieu: food and shelter. But while the bed bugs were wily on their own terms, living their secretive lives and resisting both detection and insecticide, they proved challenging to rear in the lab.

The scientists thumbed through the first sixty pages or so of Usinger's thick book for clues on how he had nurtured his own bed bugs at Berkeley. That is if they were lucky enough to find a copy. By 2007 demand for the monograph was so insistent that the Entomological Society of America, which took to calling it "The Bed Bug Bible," put out a second printing of 300 copies; these sold out, requiring a third edition within three years. But even with Usinger's work and other early scholarship, there was no easy way to start a bed bug lab. The information was old, and the expectations for such research had evolved. In Usinger's day, it was widely considered ethical and practical to regularly feed bed bugs on a person's arm, or to attach jars full of hungry bed bugs to a live rabbit like saddlebags and let the insects prick its shaved sides, or to restrain a bat on a Styrofoam block and secure containers of bed bugs along the delicate skin of its outspread wings.

While some labs still use similar methods even today, most didn't want to feed bed bugs on themselves or deal with hosting live animals. This meant they had to build artificial feeders that re-created at least the minimal cues of a human body: a temperature of around 98.6 degrees Fahrenheit and a thin skin-like membrane to trick the bugs into thinking they were pricking the flesh of a live host. Using a few papers on artificial bed bug feeders as guidance, including one published in 2002 by Spanish researchers who used the bug as an experimental model to test drugs, scientists at the University of Kentucky, Virginia Tech, and other schools started tinkering. Some built custom glass jars with tubed circulatory systems that whooshed heated water outside of the blood to keep it warm, and others bought expensive devices designed to feed lab mosquitoes. Across the mouths of these feeders, the researchers stretched various pliable sheets that could be punctured by a bed bug's slender mouth, from permeable films to thin mesh coated in liquid silicone to sausage casings to condoms.

Once the scientists had built their ghastly contraptions, they had to fill them. They tried blood siphoned from cats, cows, dogs, and horses from veterinary schools; chickens from medical research suppliers; and expired human blood that was too old for donation. But the bed bugs were picky and the supplies inconsistent. Veterinary schools often thinned their blood supplies with heparin or sodium citrate, which some experts suspected were harsh on the bugs, killing them off roughly every seven months, although no one ever proved it. And chicken breeders occasionally treated their birds with insecticides to get rid of mites; in at least two cases, the poisons seeped into the birds' blood, wiping out bed bug lab populations. Eventually, many labs settled on rabbit blood, which is readily available from medical technology companies that sell it for testing drugs due to its analogousness to human blood. Rather than using chemicals, the companies thinned the rabbit blood by mechanically removing a pulpy protein called fibrin, which spreads like frost through a blood clot and cements it in place. For a person with a cut, this is necessary; it helps stop too much blood from leaving the body. To feed a bed bug, however, it's a nuisance.

Blood with fibrin in it chunks when it's refrigerated, making it difficult for the bugs to suck it through their hypodermic mouths. To remove the protein, the companies stir fresh rabbit blood with a glass rod to initiate a blood clot. After two or three swirls, the clot sticks to the glass and is easily pulled away, allowing for up to three weeks of refrigerated storage and a nice smooth finish.

As for shelter, the bed bugs needed tight spaces kept at just the right temperature. These also needed to be portable, so the scientists could carry the insects to and from their workbenches for experiments and feedings. Most of the bugs ended up in glass or plastic jars with screw tops, altered with mesh covers. Inside, folded cardboard or filter paper provided the cozy nooks for the bugs to hide. Most labs stored rows and stacks of the jars in temperature-controlled incubators. Once, on a tour of the bed bug lab at the Ohio State University, I saw inside one of these, which looked like a refrigerator filled with jars of jam that had gone very bad. ("They like it dirty," the lab's head entomologist told me as I peered through the glass of one of the jars at hundreds of bugs huddled amongst their detritus.)

To complete their new labs, the researchers needed bed bugs to seed their collections, some of which would grow to tens of thousands of bugs representing dozens of unique strains. The scientists recruited most of these early colonists directly from infested bedrooms in cities such as New York and Cincinnati. As the pest controllers had known for some time, these bugs were showing signs of pyrethroid resistance, sometimes walking directly through piles of insecticide dusts or dried residues without so much as a shiver.

In order to figure out just how resistant these rugged bugs had become, and to decipher the genetic mutations responsible, the researchers also needed bed bug strains that were still vulnerable to pyrethroids to use as a baseline. Standard genetic analysis would make it possible to search for the underlying molecular differences between the vulnerable and the resistant bugs, which could eventually help chemical makers develop new pesticides.

The ideal candidate for this job would be a bed bug strain that had been sheltered from the bombardment of modern pesticides for a long stretch of time. One such population had been living a quiet life in a row of Mason jars in a Maryland suburb.

• • •

Up until the bed bug resurgence, the bugs raised by Harold Harlan, the army entomologist who had rescued the insects from the New Jersey barracks decades prior, were not widely known in entomological circles. That changed in the late nineties when Harlan's name started getting passed around among experts encountering bed bugs for the first time. Initially, they begged for information about what these things were and how to get rid of them. Soon Harlan was fielding calls and e-mails from academics around the country seeking advice on rearing bed bugs in the lab. Next came requests for starter batches of his homegrown strain, which he dutifully packaged into small plastic vials and mailed through the United States Postal Service, a courtesy he continues today.

The summer of 2011, more than a decade after those initial cries for help, I perched on a chair in Harlan's office in a Department of Defense annex north of Washington, DC, which was crammed with entomology textbooks and swag from pest control conferences including rubber spiders and snakes. Harlan sat in a desk chair across from me, a balding man with bushy eyebrows in his sixties. He wore wire-rim glasses and a silk dragonfly tie that was clasped to his dress shirt with a colorful pin shaped like an insect. When I admired it, he explained it was made from a real caddisfly, the larvae of which build their own armor with stones and sand glued together with silk. A jeweler he knew raised the larvae with flecks of gold and semi-precious stones and then fashioned the resulting creations into wearable art.

I watched as Harlan attached a small, flat box to his arm with a rubber band. He had modified one of its sides with fine mesh. The box was originally intended as a display case for a prized coin or medal, but as he pointed out the forty or so hungry bed bugs inside, he explained that he used it for feeding demonstrations. On a table to my right sat several Ball brand jars with mesh-topped lids sealed along the edges with layers of duct tape and rubber bands. In a more typical feeding, Harlan said, he would hold each jar against his arm or leg to nourish more bed bugs at once. I could

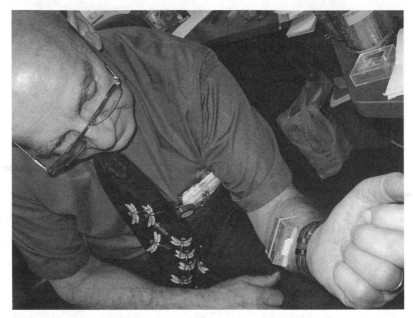

Harold Harlan feeding bed bugs on his arm. Credit: Brooke Borel.

see a handwritten label on one jar that tracked the dates of each meal.

Harlan had even more jars at his house. Collectively, they held around six thousand bugs, which he fed around once a month. By my calculations, that meant he got at least seventy-two thousand bed bug bites per year. When I asked if he ever had an allergic reaction, he rolled up a pant leg to reveal one of these bites—a bright pink welt on his calf the size of a silver dollar.

While the greedy bugs sucked Harlan's arm, I held in my hand a gift he had given me at the start of our interview. It was a black-lidded small glass vial holding about three-dozen dead bed bugs of various ages preserved in alcohol. This vial would later sit on my desk just under my computer monitor or, more often than you would like to imagine, gripped loosely in my forefingers and thumbs, where I would slowly rotate it and watch the bed bugs tumble in the clear liquid like a storm in some apocalyptic snow globe, alternately obscuring and revealing a wisp of white paper which read: "Common Bed Bug, *Cimex lectularius* L. from Reared Population/Sept. 2010 Crownsville, MD/H. J. Harlan."

As the bugs on Harlan's arm fed, I leaned in for a closer look, watching as they grew plump and turned red.

"What does that feel like?" I asked.

"It's kind of a very slight pinprick," he said. "It's hard to describe; it's sort of like someone is taking a paintbrush—you know, an artist's paintbrush, sort of a stiff one—and moving it around."

As he described the intricate mechanics of a bed bug bite, he gestured with his hands and occasionally clasped them in front of his chin as though praying as he paused to think. The small box stayed put. He glanced down and then held his arm out to me: "As you can see, some are beginning to get larger. This one over here, this male's getting pretty big."

When the bugs finished their meals, Harlan set the box aside and gestured me toward a microscope on a nearby table. He opened a container, shook out a few dead bed bugs and bat bugs into the shallow wells of a white plastic tray, slid the tray under the microscope, and fiddled with the adjustment knobs. Then he stepped back and invited me to take a look. The bed bugs and bat bugs were virtually identical, but, just as in Usinger's ink drawings, the bat bugs were hairier, with a bristly beard-like growth on their faces that extended across their tiny bodies. All of the bugs had black marks on their abdomens that looked like ink stains, which I learned were their last blood meals, visible through their thin exoskeletons. I grabbed my camera and snapped a photo through the eyepiece of the microscope. Most of the image was black, broken by a bright spotlight framing two bugs.

Near the end of my visit, as Harlan described the health and virility of his bed bugs with the puffed chest of a proud father ("I guess you are what you eat!"), he led me to a credenza on the other side of his office, which held a plastic tub, glass jars, a pair of latex gloves, rubbing alcohol, Vaseline, and forceps. These were Harlan's bed bug sorting tools, an art honed after years of doling out handfuls of bugs into small plastic jars to mail to new labs.

He showed me how it is done, pounding a Mason jar on the credenza in a burst of staccato thumps to knock the bugs to the bottom. As though choreographed, he tucked his tie into his shirt, snapped on a pair of gloves, and wiped the jar with alcohol. Hovering over the plastic tub, he unscrewed the lid of the jar and smeared

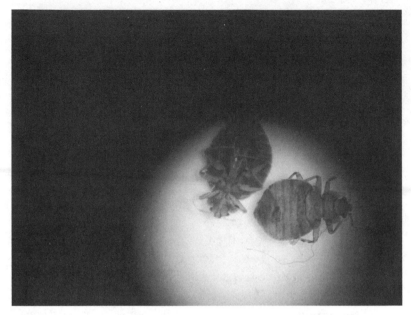

Cimicidae under the microscope in Harold Harlan's office. Credit: Brooke Borel.

a generous layer of Vaseline around its mouth, an oily barrier for would-be escape artists. With the forceps, he carefully drew a piece of cardboard from the jar and, like a jeweler selecting a set of fine diamonds, plucked ten bed bugs, dropped them in a small jar with a cardboard strip inside, twisted on the lid, and handed it to me. I held it inches from my face and watched the bugs as they climbed across the cardboard. I brought the lid to my mouth and exhaled across the mesh top, and the bed bugs scrambled upward, alerted by the carbon dioxide from my breath. Faint vibrations tickled my fingers when the bugs moved, although it didn't seem possible that the sensation was real. I held the jar farther away, pinching it between my thumb and forefinger. The bugs shifted from the lid to the points where my skin met the glass, attracted to the heat radiating from my body. Unsettled that they were just as interested in me as I was in them, if not more, I set the vial back on the credenza.

It was as clear to those first researchers trying to raise bed bugs in their labs as it was to me when I sat in his office that day: Harlan's bugs were no longer the odd hobby of an eccentric army entomologist. For decades, he had raised his bugs in isolation. They had lived a steady life, away from pesticides and other trauma for nearly forty years, and so they had all the hallmarks of a good comparison strain. They were susceptible to all pesticides, they were robust and generative, and, unlike the bed bugs captured from apartments and houses, they were easy to rear. They took to artificial feeders like a baby to the bottle. Perhaps generations of coddling and consistency had left them more accepting of this approximation of natural life. Harold Harlan had unintentionally created just what the bed bug labs needed.

• • •

Thanks to the persnickety nature of the bugs collected from people's homes, it took a year or two for the labs at the University of Kentucky and Virginia Tech to prepare for the real research, which they hoped would show how the bugs were spreading so suddenly and why they were so damn hard to kill. Experiments over the next several years came to fruition in a series of discouraging papers published in 2006. Virginia Tech showed that strains from an Arlington apartment complex took nearly 340 times longer to die from a common pyrethroid than the Harlan strain. The University of Kentucky found bed bugs in Cincinnati and Lexington that were collectively 12,765 times more resistant to pyrethroids, also compared to Harlan's bugs.

Experts overseas were also finding high levels of resistance. Researchers in London discovered that bed bugs from private homes and hotels around the UK had virtually no response to the concentration of pyrethroids that killed 99 percent of both Harlan's bugs and two other susceptible strains, including one that had been raised in Monheim am Rhein, Germany, at the chemical company Bayer CropScience for several decades. (While Harlan's bugs were unusual, they were not entirely unique: similar bugs lived in a handful of labs across the world, relics of old insecticide programs.) And in Australia, entomologists compared the vulnerable German bed bugs to common bed bugs collected in Sydney and found that the latter were more than 400,000 times more resistant to a pyrethroid called deltamethrin and a whopping 1.4 million times to another, permethrin.

The bed bugs were clearly very resistant, and their hardiness likely contributed to their spread. If a bug is difficult to kill, it might survive long enough to hitchhike to a new home, and it will also pass its toughness to new generations. But how did the resistance work? In 2008 the issue came to the attention of scientists at the University of Massachusetts Amherst who had been studying the genetic basis of pyrethroid resistance in lice that was likely driven by topical insecticides in shampoos and creams. Curious whether the same genes were at play in the bed bugs, the researchers tried a series of experiments on two groups of the insects. One was a highly resistant field strain from New York. The other was reared in a Florida lab for more than twenty years and, like both the Harlan and Monheim bugs, was one of the world's rare susceptible strains. The Florida bugs may have been the descendants of the bed bugs the Aberdeen researchers used in their Vietcong insect detector experiments in the late sixties, although the insects have passed down from researcher to researcher to the point that no one knows for sure.

The UMass team first ran a standard experiment like those from Virginia, Kentucky, the UK, and Australia, which showed that the New York City bed bugs were 264 times more pyrethroid-resistant than their sheltered Florida brethren. Then the scientists looked at the specific set of genes related to pyrethroid resistance in lice. Genes furnish instructions for making proteins, which in turn provide the structure of a living being's body as well as the functions within it. All individuals within a species have the same set of genes—that's one of the things that distinguishes a bed bug from a human from any other organism, genetically speaking—but each will have slight variations that give them their unique characteristics. These may include pyrethroid resistance in a bed bug, for example, or hazel eyes in a person.

These variations are caused by mutations, or changes in how genes are built. The genes that the UMass scientists were interested in are responsible for programming sodium channels, which are tube-shaped proteins in the membranes of cells, including at the end of a nerve cell. These proteins are essentially tiny molecular machines that allow sodium to pass through the membrane and into the cell, which it does as one of many necessary steps in a nerve impulse. In the nervous system of a regular insect—indeed, most animals—signals travel from the brain to the rest of the body and back across long, stringy nerves like voices zipping across a telephone wire, telling the body how to interact with and react to the world. Rather than electricity traveling down this biological wiring, the signals take the form of a cascade of electrically charged salts. To make it from nerve cell to nerve cell so the messages jump across the body, gates on one end of the tubes swing open to let one type of salt, sodium, pass through and shut when their role in the message is over. (The system requires another kind of salt called potassium, too, and both salts are pumped by a different molecular machine to reset the nerve so it is ready for the next message.)

When a Harlan bug is dosed with a pyrethroid, the insecticide makes its way to the nervous system and clings to those tube-like sodium channel proteins, essentially holding the gates open so that when nerves start sending messages, they can't stop. This is why the bug convulses to death. Pyrethroid-resistant insects have a specific mutation to the relevant genes, which makes the genes build a slightly different sodium channel protein, shaped so the pyrethroid can't get a good-enough grip to hold open the gates. The signals stop when they're supposed to, and the bed bug lives.

Since the UMass scientists already knew which genes controlled the sodium channels in lice, and what the mutations that caused the resistance looked like, they just had to test the same genes in both the New York and Florida bed bug strains. Specifically, they had to read the order of the nucleotides——essentially the letters in the genetic alphabet—in the strands of DNA that made up these genes and then compare them. What they found was this: the resistant bugs had two single-point mutations compared to the susceptible bugs, either of which might cause the resistance. In other words, if the genes were sentences that described the sodium channels, the New York and the Florida bed bugs were only different by two mistyped letters, and either of those typos might be responsible for the pyrethroid resistance.

In insects, these sodium channel mutations are called knockdown resistance, and they are common in more than just bed bugs and lice. The name comes from an observation of DDT-resistant house flies in the 1940s. The knockdown effect simply described how flies exposed to the insecticide had trouble flying, or even walking, which made the insects easy to knock down. A knock-down genetic mutation decreases an insect's nerve sensitivity to a particular pesticide. It is the same type of resistance that researchers found in bed bugs and a slew of other insects in the decades following World War II, as well as the same cross-resistance discovered between DDT and pyrethroids. This makes sense: if both DDT and pyrethroids attack the same part of the nervous system, it follows that a physical shift in the sodium channel that thwarts one would similarly stop the other. Bed bugs consistently bombarded with DDT would have maintained these mutations. And while the resistance may have faded in the bugs that caught a break from DDT after its ban, the fact that their recent ancestors had the resistant genetic information allowed it to spring back faster once they were exposed to pyrethroids.

Once scientists had identified what likely made the bed bugs so resistant, they tried to see how widespread the mutant bed bugs were. Again it was grim. In 2010 the Kentucky researchers published a paper showing that about 88 percent of more than a hundred bed bug populations collected from Kentucky, Ohio, Michigan, New Jersey, Massachusetts, and New York had one or both of the mutations that made them so resistant; in an unpublished study the same year, scientists from North Carolina State University got virtually the same results from bed bugs in fifteen states, including California, Nevada, Texas, and throughout the Midwest and up the East Coast.

Some of the bugs, however, were unyielding to pyrethroids even without the knockdown mutations. Another team from the University of Kentucky, led by an insect physiologist and toxicologist named Subba Reddy Palli, learned that bed bugs sometimes had super metabolisms that were especially efficient at dismantling the poisons. These bugs had evolved improved versions of proteins called enzymes, which chopped up the pyrethroids and passed them harmlessly through the bugs' bodies like a sot who is able to guzzle more and more booze without getting drunk. The researchers discovered nearly eighty of these pyrethroid-busting enzymes, with more likely to come, and it was possible that the altered enzymes were also driven by genetic changes. Still other research from Virginia Tech suggested that bed bugs might be growing thicker exoskeletons, which could block insecticide from working its way into the nervous system.

Even worse, the bed bugs could have a combination of all of these alarming new traits. By 2012 scientists had found strains with both knockdown mutations and amped enzymes, and it wasn't too much of a leap to think that there might be some bugs with one or both of these plus the thicker shells, or with some other yet-undiscovered mode of resistance written in the bed bug's genes.

• • •

One way to find genetic mutations in a living thing is to look at its genome, or the complete set of its genetic information. In 2011 the Bed Bug Genome Consortium, a collaboration of roughly two dozen researchers, aimed to do just that with the bed bug. The research was part of a larger project called i5K that hopes to uncover the genomes of 5,000 insect and arthropod species in five years, sometimes called the "Manhattan Project of Entomology." For the bed bug, the goal was to determine the precise arrangement, or sequence, of the alphabet that spells out the insect's two genomes: its nuclear genome, which consists of the genetic material found in the nucleus of nearly all cells in the body and is inherited from both the mother and father, and the mitochondrial genome, which includes the unique genetic material belonging to little energy-producing cellular furnaces called mitochondria. Mitochondrial DNA is passed from one generation to the next via the maternal line; only traces of the mother and her mother and her mother's mother, and so on, are found within it, and those traces remain identical from one generation to the next. A list of all base pairs that spell out both the nuclear and the much smaller mitochondrial bed bug genomes would provide the genetic template of the species, describing its tens of thousands of genes. (Base pairs are the double strands of nucleotides, the genetic alphabet, which form the basic structure of DNA's famous twisting double helix.)

But having a list of genes is not enough to understand how a living being functions. Genes are an instruction manual that requires several stages of reading and copying before it can be translated into the proteins that make up the structure and function of a working body. In one of several important steps, a molecule called RNA—which in a way is a copy of the DNA—delivers the

DNA's message to the cellular department that manufactures the proteins. A collection of all the RNA at any given time is called the transcriptome, which gives a snapshot of what is happening in a body, or what proteins are being made, which changes from one moment to the next.

To read the bed bug's genome, the consortium used upgraded versions of the sequencing machines that the Human Genome Project employed a decade earlier. Gene sequencing is a painstaking process in which scientists must break up DNA into small chunks and feed them one by one into the machines, which then spit out the base-pair lettering of each chunk. By early 2013, the consortium had stitched together all 960 million base pairs that make up the bed bug genome, roughly a third of the number in the human genome. It's estimated that less than 5 percent of the bed bug's hundreds of millions of base pairs code for the transcriptome, the collection of RNA, at any given time. Within the genome, the researchers surmised, there were likely around 20,000 functional genes (humans have around 25,000).

By the project's end, researchers worldwide will have annotated specific regions of the genome that represent specific genetic roles. The information will be available online so that anyone who wants to see what makes a bed bug genetically a bed bug can do so. Perusing the bug's genetic story might reveal new pesticide-resistant mutations, or give insight into how female bed bugs are able to survive traumatic insemination, or show how bed bugs know to hide in tight near-invisible spaces; any of these could help point to a new way to kill them.

The consortium isn't the only team working toward this goal. One night at a Brooklyn bar, I learned that a neighbor of mine, an evolutionary geneticist, is also working on the bed bug genome. After some initial confusion where I thought he was part of the Bed Bug Consortium team, I realized he was in a rival group of scientists, although at the time neither group knew of the other's intentions. Through collaborations with the American Museum of Natural History and Cornell University, his team at Fordham University is using identical methods and machines as the Bed Bug Genome Consortium and will finish roughly at the same time with aims to publish in the same highbrow scientific journals.

Both of the teams are also using the same kind of bed bug. A single bed bug doesn't have enough information for the sequencing machines to read. It's just too small. Capturing the genome requires a larger volume of genetic soup from hundreds of bugs. And those hundreds of bugs need to be as genetically similar as possible, so that the soup will be uniform and thus easier to match strips of machine-read DNA with one another to reconstruct the genome. For those researchers interested in finding new ways to kill bed bugs, it is also important for the bugs in the study to have as few genetic mutations for pesticide resistance as possible, so that resistant strains can later be sequenced and their mutations will stand out against this original, uniform, susceptible genome. The perfect subject for both bed bug genome groups is the inbred and chemically vulnerable Harlan strain, which means that the bed bugs Harold Harlan collected decades ago will be the first strain in the world to have its genomes sequenced.

• • •

By late 2012, just two weeks after I watched the cast of *BEDBUGS!!!* dance across the stage at the Chernuchin Theatre, I became acutely aware how deeply immersed I was in the world of bed bug research. So deeply, in fact, that I had scraped together money to go to Knoxville, Tennessee, for the sixtieth annual meeting of the Entomological Society of America, a trip I couldn't have fathomed as vital even a year earlier when I sat in Harold Harlan's office and watched him feed bed bugs on his arm. The Knoxville meeting showcased the work of hundreds of entomologists for thousands of participants, and bed bugs had become such a hot research topic that there was an entire special section dedicated to the insect: "Bed Bug Research: Catching Up with the Global Bed Bug Resurgence." It was four hours long, and I was in bliss.

During one of the bed bug talks, I learned that population geneticists had been quietly gathering specimens for a project that might complement the bed bug genome work. Population genetics is the study of gene flow—the distribution of different versions of genes called alleles throughout a species across space, time, or both. This information helps determine how different members of a species are related and how they've migrated and evolved, and it can be mapped on a location-based family tree called a geophylogeny. With bed bugs, such data might help answer these questions: When and where did the resistance first start? Were the bed bugs in the United States homegrown? Or had the resistance spontaneously emerged worldwide virtually everywhere at once, or at a bed bug epicenter from which the fallout slowly spread across the globe? At the conference, a soft-spoken molecular ecologist from North Carolina State University named Ed Vargo flipped through a PowerPoint presentation describing how he might answer these questions by tracing the flow of bed bug genes collected from homes and museums from around the world, snapshots of the bed bug's genetic history both in the present and in the past. At one slide, I stopped my rapid tapping on my laptop, which was balanced precariously on my lap, and stared. Some of the bed bugs had come from Usinger's collection at Berkeley, where I had become smitten with the stories of the bed bugs gathered from all over the world.

Months later over a series of several phone calls, I would ask Vargo, who seemed a bit bewildered at my rabid curiosity for the topic, for every detail of his visit to the Essig Museum and its aftermath. It went something like this. On a brisk spring day in 2009, Vargo and his postdoctoral student, a Northern Irishman named Warren Booth, scratched their names in the guest book at Essig, just as I did two years later. The scientists had received permission to sample DNA from some of Usinger's decades-old bed bug collection to include in their global population study.

The museum curator presented Vargo and Booth with a different set of bed bugs than were offered to me during my visit; instead of shelves of black books holding hundreds of paper-thin bugs on glass slides, the scientists were given a drawer lined with open boxes. Tucked inside were sixteen large glass jars, each holding a dozen glass test tubes sealed with cotton and stoppers. The tubes held whole bed bugs, sometimes a handful and sometimes hundreds, floating in alcohol. These were better candidates for DNA extraction than a dried bug smashed on a piece of glass, although their stories were not so well kept. The labels on the vials sometimes said no more than the year, the species, and the general area where it was collected, as unspecific as "California" or "Czechoslovakia."

Vargo and Booth had permission to take two specimens from each tube, but there was a catch: they had to return the bugs virtually unscathed. And so they pulled on nitrile lab gloves and began to unplug the test tubes. They deftly plucked the cotton filler and carefully pinched with their forceps local bed bugs from Berkeley as well as samples from Ohio, Cairo, Nagasaki, and Pittsburgh, along with three other related species from around the world. Each went into a plastic screw-top vial filled with alcohol, on which the researchers copied each scant label in pencil. Then, as though they were checking out books from a library, they signed their names on a stack of paperwork, dropped the vials in a bag, and took them home to North Carolina.

Through another series of intense and detailed phone calls with Booth, I would learn that extracting DNA without disturbing the integrity of the bed bugs had required delicate treatment. Back in the lab in Raleigh, Booth removed the fragile insects from the vials one by one, held them steady with a pair of tweezers in the tray of a dissecting microscope, and took a finely bladed scalpel to a seam on their segmented underbellies with the precision and aesthetic concern of a cosmetic surgeon. He placed the bugs in a liquid detergent; for several days, the fluids and soft tissues inside the bugs bled into the clear liquid, which broke down the cell walls to release the mitochondria, the nucleus, and other organelles within (organelles are distinct cell parts—the organs of a cell). He coaxed out the genetic material inside using enzymes to further chop up the organelles. A turn in centrifuges separated those pieces from other cell parts like a washing machine on the spin cycle pulls water from freshly cleaned clothes. Alcohol helped free the DNA within, and more spinning separated the DNA from the rest of the cellular material. Each step stripped away the unwanted molecules through chemical reactions and mechanical sifting until all that was left was DNA.

Because the Usinger collection was so old and the whole bugs were poorly preserved, their DNA had degraded to the point that Booth could only pull fragments of genetic information. From these, he could only fill in a small part the geographical family tree. The only tool at his disposal was to look for the presence or absence of the two major knockdown mutations. Combined with the year the bug was collected, the knockdown data can show patterns of pyrethroid and DDT resistance. The results were intriguing but didn't tell a complete story: the oldest bugs had one, but not both, of the mutations, suggesting that at least one had sprung up after the introduction of DDT. If the scientists could find even older bugs, from the forties or earlier, they might be able to show a deeper connection between the pesticide and the resistance.

Around the same time that Vargo and Booth were working with the Essig collection, they had mapped around two hundred specimens of modern bed bugs caught from twenty-one apartment complexes along the East Coast of the United States. This time, since the genetic material was fresh, the researchers could use two additional analytic tools: DNA genotyping and mitochondrial DNA analysis. DNA genotyping looks for genetic markers called microsatellites, which are regions in the genome where base pairs are repeated. Both the father and mother pass on these repeated sequences, or slight variations of them, to their offspring. This means that the microsatellites can identify individual bugs and show how they relate to all the other bed bugs around them, which is particularly handy for understanding the familial relationships of bugs living in a single room or home, or dispersed across larger areas. Because mitochondrial DNA shows genetic characteristics passed along the maternal line, which are identical for the females of a family stretching back across all generations, it clarifies how bed bug families relate over time.

The results were perplexing. The map had an unexpected structure. The nuclear DNA showed bed bugs from two buildings in the same city as wildly different, while the overall genetic diversity in one city versus another was pretty much the same. In other words, the bed bugs in each city seemed to be coming in from multiple sources, which matched the hypothesis that bed bugs came from many places rather than concentrated populations in nearby poultry farms or housing complexes. And since there was no clear distinction between genetic variants found in Ohio versus, say, New York, those populations hadn't been separated for very long. They likely had a fairly recent common ancestor. Adding to the curious patterns was the genetic distribution of infestations within a single apartment complex, which were incredibly closely related;

so much, in fact, that the data suggested a widespread infestation covering multiple units could erupt from a single gravid female, who could go on to mate with her children, who would also inbreed amongst themselves.

The fledgling family trees were interesting, but the scientists didn't have enough data to fully explain the bed bug's movement. That would require far more bugs from around the world. Still, the work led most bed bug researchers to agree that the resistant insects had come from many sources. But as I asked scientist after scientist where, specifically, the bugs had come from, I realized they couldn't agree on much else. Members of the Ohio State University bed bug team suggested that their state's infestations were due to an influx of Somali immigrants, based on the high number of bed bugs found in those communities in Columbus and Cincinnati. In Kentucky, the first bed bugs during the resurgence were found in the homes of Sudanese refugees known as the Lost Boys, which made Potter and others wonder whether these bed bugs, too, had come from Africa. Overseas, it was no different. In England, one expert told me that some people blamed immigrants from Kenya, others said the bed bugs were imported by fellow Brits who took holidays in Turkey, and still others claimed the source was a piece of luggage from Australia. Or, perhaps, it was Brazilian sightseers, although the same expert also commented that he had colleagues in South America who pointed the blame back to the United States. Brazilian tourists, they said, were allegedly bringing the bugs back from Disney World.

The population geneticists from North Carolina saw a different possibility. To Vargo and Booth, as well as their colleague Coby Schal, an expert on insect behavior, it didn't make sense that the pyrethroid resistance had simultaneously and spontaneously appeared worldwide. Instead, they argued, it likely began in just one or two spots. From there, the hardy bed bugs radiated, getting transported from one place to another through the increase in travel. In the globalized world, it wasn't too hard to imagine the bugs hitchhiking first from their origin to a few key international hub cities and then being whisked away and scattered worldwide where they quickly multiplied and spread.

The origin, wherever it was, would have to be in a temperate or subtropical region where the common bed bug thrived, and it would had to have been doused with pyrethroids or DDT, not just for cockroaches in kitchens and bathrooms, but also in bedrooms. It would also have to be a place that had people traveling in and out in the years leading up to the resurgence. And bed bugs there would also need to have the genetic signature of the bugs found throughout Europe, the United States, and other parts of the world with new infestations.

While pyrethroids had been used globally for decades, they were especially common in malaria-ridden regions including Southeast Asia and parts of Africa. The pesticides were not only regularly sprayed in bedrooms to protect people from mosquitoes as they slept, but also impregnated in the netting draped over their beds. This constant presence would artificially select the bed bugs that had to wade through these poisoned obstacles each night to feed, allowing resistant insects to thrive. But Southeast Asia was mainly home to the tropical bed bug, a species so far rare in the US resurgence. Africa, however, had both tropical and common bed bugs, and there was overlap between the latter and regions heavy with malaria, including Kenya, Somalia, and Sudan. The idea had some experimental basis, albeit thin. Research from the late eighties showed that bed nets dipped in a pyrethroid controlled tropical bed bug infestations in Gambia and across the continent in Tanzania in 1999, while later work from 2002 showed that the bed nets were forcing resistance in the species in Tanzania. But so far, the few bugs that Vargo and Booth had studied, which were collected around 2000 in malaria-ridden Zambia, didn't show the same patterns of knockout mutations found in Europe or the United States.

Another possibility, they reasoned, was Eastern Europe. Pyrethroids had been used in the region for decades. Anecdotal stories suggested that bed bugs had been common there even before their global resurgence, including in Russia, where the poet Mayakovsky had penned his bed bug play decades before, and the Czech Republic and Slovakia, where Robert Usinger had planned to travel to study yet more bed bugs before his death. Bed bug samples sent to the North Carolina State University lab from a colleague in Prague, a doctoral student named Ondřej Balvín, suggested that the bugs found throughout Western Europe and the United States may be genetic subsets of those found in Eastern Europe, although the data wasn't strong enough to be sure. It was intriguing, though. When the Berlin Wall fell in 1989 and the USSR dissolved two years later, the region was more open to travel both coming and going— the timing reasonably poised the bed bug to spill over to the rest of the world in the following decade.

ANNIHILATION
By Any Means Necessary

One cold October night back in 2009, just before 3:30 in the morning, a fire truck raced toward a quiet suburb northwest of Cincinnati. The firefighters arrived to find flames shooting from a second-floor window in a brick apartment building. They jumped from the truck and ran to evacuate the dozens of people who lived in the apartments, who then shivered in the cold night air and watched as the fire licked at a tree growing outside the burning unit. The firefighters pushed their way toward the source of the flames. Smoke billowed from a bedroom; inside, they found a burning mattress and box spring leaning against the wall. The fire was out twenty minutes after their arrival. Based on the charred evidence and some interviews, the team pieced together that a young man who lived in the apartment had doused his bed with isopropyl alcohol to try to kill bed bugs. The official story was that the man had then lit a cigarette and the spark from the lighter flashed into a ball of fire. No one was seriously hurt, but there was at least $20,000 in damage and the building manager kicked the man and his family out of the apartment.

The unofficial story was a little different. When I first read about the fire in an archived Ohio newspaper, I couldn't stop thinking about it. Could bed bugs truly trigger a person to do such a thing? What kind of scenario leads someone to spray alcohol all over a room and then light a flame? And could I have ever been driven to the same desperate conclusion? I went back to the article and saw that it mentioned the apartment building's street name but no address. I had to find where this happened. I mapped the street and city and zoomed in with Google's dynamic street view. I moved in and out, scanning the length of several blocks for an apartment complex. There. I zoomed back in and looked at it from every possible angle. The brick building sat among neat green yards. It revealed its street number but no other clues. Soon I was scouring the Internet for the exact address, trying to find contact information or any other tie to an actual person who could tell me what happened that cold autumn night in this nondescript place. I found a unit from the building advertised on a rental website—no mention of its buggy past—and I sent a message through the automated contact form, figuring it would go unanswered.

Not long after, my phone rang. It was the building's manager. After some initial confusion about what I was after and, I sensed, some disappointment that I wasn't a potential renter, she told me her version of the story, which was nearly the same as what I'd read in the paper. Except, she said, it wasn't cigarettes.

"He sprayed that room and lit his bong. And it shot like a cannon across the room. All the alcohol fumes—everything just went up."

"A bong?" I asked. "You mean he was smoking pot?"

"Yes," she said. "He was baking a cake, too. Oven was on and everything. Who does that kind of thing at three o'clock in the morning?"

It turned out that getting high wasn't the only way to drive a person to light their bed bugs on fire. As I searched through more archived stories, I learned that people all over the country had been accidentally setting their houses on fire while trying do-it-yourself bed bug eradication techniques—lighting cigarettes after spraying their bed or couch with rubbing alcohol, flammable over-the-counter pesticides, or a mixture of both. Other home owners set fires by using faulty space heaters or radiators, which they had cranked on high after hearing that heat kills the bugs. This usually happened when people tried to avoid paying thousands of dollars for a professional heat treatment; later I'd learn that not only had experts also set house fires, but there was even a special business insurance to cover them.

I learned about poisonings from bleach, chlorine gas, and over-the-counter pesticides meant for the garden. Experts told me about people setting off too many bug bombs simultaneously, like a sad fireworks display, despite evidence that the products didn't even kill bed bugs. And in a 2011 "Morbidity and Mortality Weekly

Report" from the Centers for Disease Control and Prevention, I read about a North Carolina man who accidentally killed his sixty-five-year-old wife—who already suffered from kidney failure, a bad heart, type II diabetes, and depression—after treating their home with two unregistered pesticides and nine cans of fogger.

Not everyone was shocked by these cases. When I asked Matt Beal of the Ohio Department of Agriculture about the 2009 Cincinnati blaze, he said he had not been surprised to hear about it. At the time of the fire, he had been reading stories of people accidentally burning their homes, and worse, for months in the state, as well as across the country. Partly because of these desperate self-treatments, and partly at the urging of the Ohio Pest Management Association and a long acronymic list of other pest control organizations, Beal had been preparing for months to ask the Environmental Protection Agency for help with his state's bed bug problem. As Beal and his allies saw it, that problem was rooted in the fact that there was no cheap, effective way to kill the bugs, particularly in places where a lot of people lived tightly packed in the same building: apartment complexes, public housing, college dorms, retirement communities, and homeless shelters. A single infestation in these structures is a hydra: knock off the bed bugs in one unit, and they emerge as vile and virulent as ever with the neighbors.

To control infestations, exterminators used—and still use today—several tricks together in an approach called integrated pest management. In addition to insecticides, this may include cleaning and vacuuming; heat treatments (which bake the bugs until they resemble flakes of red pepper); laundering clothes and bedding in hot water and drying them on high; blasting furniture with industrial steamers; sprinkling dusts around the room, either made of crushed fossilized algae called diatomaceous earth or silica, both of which absorb the fats in an insect's exoskeleton and dehydrate it to death; and fumigation with sulfuryl fluoride, which involves pumping a tented building or a trailer full of a poisonous gas. But many experts, especially those who are friendly with the chemical industry—which is nearly all of them—maintain that the most important affordable option for the nastiest infestations is insecticide. And thanks to resistance, pyrethroids work too slowly to master intractable infestations, if they work at all. The Ohio exterminators needed a better exterminator.

Beal had seen data from Mike Potter and his colleagues at the University of Kentucky that compared the available pyrethroid products to other pesticides, most of which could be used on bed bugs in the lab but couldn't legally be sprayed in a bedroom. One, a carbamate called propoxur, common to flea collars, killed every bed bug it touched. The pesticide had been available for residential use just two years earlier, which meant that the EPA once thought it was safe enough to spray in homes, or at least that it posed no unacceptable risk. (The agency skirts the word "safe" in part because no pesticide is entirely harmless at all doses and in all possible scenarios.) But the pesticide was soon in an administrative gridlock.

In order for any pesticide to make it to the American market, the company selling that pesticide has to prove its relative safety by performing extensive toxicity and environmental tests. The EPA reviews the resulting data and either approves or denies the application. If approved, the EPA dictates where and how it can be used and registers a detailed label outlining those restrictions. Follow the label as it is written and the pesticide won't harm you, at least according to the scientific research that went into testing it.

Once a company has a registered label, it has a few options. One is to manufacture and sell the product. Another is to license the technology to a different chemical company that then manufactures and sells it. A third is to sell the registration to another company, transfer ownership of the labels to that company, and have nothing to do with the product.

In addition to labeling, the EPA also has the authority to question a pesticide's appropriate use if new data show that the chemical may no longer be reasonably safe in a particular setting. In 2007, although there wasn't evidence that propoxur posed an unacceptable risk, the EPA decided it wanted more proof that the insecticide passed new safety standards intended to protect young children, who, the agency theorized, might run their hands across a propoxur-treated baseboard and then shove those hands in their mouths. The EPA told the major chemical companies that owned the propoxur registrations they would need to run new tests to prove it was reasonably safe to use in homes. Coincidentally, those companies were in the middle of selling the registrations to yet another set of companies, and the new toxicity data cost too much for a product they weren't keeping. They simply canceled the residential registration. After all, there were other markets. The pesticide could still be used in office buildings, meatpacking plants, and other places where children usually didn't go.

Since there was no definitive data that propoxur actually posed a threat to children and the pesticide had so recently been used in homes with no apparent problem, Beal petitioned the EPA for a Section 18 Exemption, named for the relevant section of Federal Insecticide, Fungicide, and Rodenticide Act, which allows the EPA to control pesticides in America. Under this exemption, the EPA can let a state use a pesticide outside of the chemical's registration, so long as there is an emergency that warrants it and the pesticide is reasonably safe for the requested use. Armed with five letters of support from university entomologists and chemical company representatives, Beal submitted a petition for three propoxur products just a week after the bong fire broke out in the Cincinnati suburbs.

Then he waited.

. . .

Early in my bed bug quest, even before I knew about people catching their houses on fire or poisoning themselves with insecticides, I traveled to Virginia to see Dini Miller, the entomologist from Virginia Tech who was among the first in the country to set up a modern bed bug lab. One evening during my short visit, we sat in a booth at a Thai restaurant, and Miller rapidly chronicled the drama of bed bugs and public housing in her state, an area where she focused much of her research and energy. I scribbled into a notebook in between bites of curry. The laws, she said, simply weren't good enough when it came to disclosing bed bug infestations or helping control the insect's spread. The conversation shifted to insecticides and how they are regulated, which directly affects how pest controllers can treat an infestation. It was here that I first heard of propoxur and the stalemate between Ohio and the Environmental Protection Agency.

"It's directly related to the whole risk cup thing," she told me.

I stopped writing and looked up at her. "What? Risk cup? Like an actual cup?"

She looked back at me and sighed, which read, to me, as a combination of pity and resignation. It wasn't an easy topic, and her look seemed to say, *Lady, you do not know what you're getting yourself into.* Then she gave me a crash course in the EPA's risk cup concept, using the glasses of Thai iced tea and water that sat on our table as educational props.

I nodded as Miller told me about imaginary cups and how they measure out insecticide exposures, but I didn't entirely understand even after her lecture. Many months later, it took days of sifting through regulatory agency documents and a painfully detailed hour-long phone interview with the EPA to truly grasp the risk cup. So here it goes. Picture for a moment a small cup. It is clear plastic and graduated like the lid on a cough syrup bottle. Over your lifetime, say seventy years, every time you get a small dose of insecticide after eating an apple you didn't wash off, or going a little overboard with the Raid after a roach scurries under the bathroom sink, or throwing a ball with your propoxur-flea-collar-wearing dog in the backyard that you recently treated with herbicide, the traces of each pesticide go into that cup. Theoretically, as long as the cup isn't full, it isn't likely that those pesticides will give you cancer or otherwise threaten your health. Too much exposure and the cup overflows.

The risk cup idea has been around for thirty or so years, but its modern form debuted with the passage of the Food Quality Protection Act of 1996, an amendment both to the Federal Insecticide, Fungicide, and Rodenticide Act and the related Federal Food, Drug, and Cosmetic Act. The Clinton administration ushered in the new food act, which included new pesticide safety requirements that had been upgraded with the newest science available, partly inspired by Rachel Carson and her *Silent Spring* legacy. The government wanted to make sure common pesticides didn't pose a risk to human health and particularly the health of children.

Before the new food quality rules, the federal government essentially had several risk cups for each agricultural pesticide from each of the possible places a person might encounter it, including food, water, and around the home or in recreational settings, such as parks and golf courses. In order to determine the risk of the chemicals, the EPA required toxicity data from lab animals. Then they set the acceptable level of daily human exposure at one-hundredth the maximum amount that showed no effect on the animals.

With the new rules, each type of chemical had just one cup, which collected the aggregate of all potential sources of the chemical. For example, if a pesticide was used on crops in places where it could enter the water supply, or in the home or other residential areas such as on lawns or pets, the contents of its risk cup included the residues from all of those locations. The new rules also considered the cumulative impact of each group of pesticides based on how they interact with the nervous system, dedicating a separate risk cup to each group. The aggregate risk cup for each individual chemical that glommed on to the gate of a sodium channel to disrupt the nervous system, for example, all went into the same cumulative risk cup. The EPA still required animal toxicity data for all of the pesticides, but decreased the tolerance levels by another tenfold, in part to better consider children's vulnerable developing nervous systems.

As long as a specific risk cup isn't full, the EPA will consider registering new uses for a particular class of pesticide. When the cup brims, however, a company must provide new data proving that the pesticide has lower toxicity than previously shown. With the new rules, the EPA could also make any company with a product already on the market provide new toxicity data to satisfy the stricter safety requirements. This is what determined propoxur's fate. Beginning in 1997, the EPA dashed off letters to these companies, aiming first for carbamates and organophosphates because of mounting evidence of their impact on human health. Both insecticides are nerve agents. Like DDT, they attack insects' nervous systems, but their mode of action is different: rather than holding open ion channel gates to force too many nerve firings, they suppress production of cholinesterases, which are proteins that help maintain chemicals called neurotransmitters that are necessary for nerves to communicate. The end result is the same: the insects uncontrollably twitch to their death. The trouble is, we also have cholinesterase in our brains, and the EPA wanted proof that the pesticides didn't similarly damage the human nervous system.

Some of the companies that received the EPA's letters ran the tests and provided the required information, which meant blasting more lab animals with pesticides and recording the levels that separated healthy from sick, while others voluntarily let their registrations expire. In the end, many organophosphates and carbamates followed DDT into extinction in the United States and many other parts of the world, at least for residential use, effectively banning them from homes and especially bedrooms.

• • •

In 2013 on a cold January day at the Ohio State University's bed bug laboratory, housed in a plain white building on an idyllic patch of farmland in Columbus, I sat down with Matt Beal at a large conference table as he told his propoxur story in a measured, gravelly voice. More than three years had passed since he had mailed his Section 18 exemption request to the EPA. He leafed through a stack of correspondence related to the case as we spoke, pausing on key pages. The EPA's first letter was relatively deep in the pile; it had taken more than eight months to arrive, and even then it was likely prompted by an earnest letter from Ohio's then-governor, Ted Strickland, following up on the request. The EPA wrote, among other things, of its hesitations: "The specific exposure scenarios that are of most concern involve inhalation risk and also hand-to-mouth behaviors on the part of children." Then there were two more letters from the governor, more pushback from the EPA, and then radio silence. (Weeks after my meeting with Beal, an EPA representative would tell me that the pesticide was in "regulatory purgatory" and would stay there until Ohio or its partners provided new toxicity data.)

"But, then, what next?" I asked, when Beal mentioned that another twenty-plus states had expressed interest in submitting propoxur exemption requests, should Ohio succeed. "If more people are using it, resistance might happen more quickly."

Beal conceded it was possible. But, he claimed, it was unlikely because only licensed pest controllers would be able to buy propoxur. This meant it wouldn't be as overused as DDT or pyrethroids, with their numerous over-the-counter products leaving low-level persistent residues nearly everywhere. Beal told me that constant exposure to a pesticide at amounts that are too low to actually kill anything is an especially good way to drive resistance, whereas a quick blow of something far more lethal may significantly reduce an insect population before it is able to reproduce and spread resistant genes.

When I ran that idea past a couple of evolutionary biologists, they told me it is also possible that a few bugs would be naturally resistant and would have no competition for reproduction once their brothers and sisters with the susceptible genes were dead, which could lead to a balloon in resistant bed bug populations. Other scientists took a stronger stance. Stephen Doggett, a medical entomologist from Australia, told me that Ohio's propoxur move was "idiotic" and "dumb," in part because carbamate resistance had already been observed in bed bugs in both Australia and the UK. What was the point, he asked me, in going through the effort to reinstate an old pesticide that would follow the same path? Still, Beal argued that propoxur would let Ohio knock down the bed bug reservoirs in public housing, apartment complexes, and nursing homes, at least until the chemical companies came up with a new product, which could go into rotation with propoxur while pyrethroids took a hiatus.

Already there were new bed bug products coming to market. One was a combination of pyrethroids and neonicotinoids, a synthetic version of nicotine that may be partly to blame for the mysterious decline of the honey bee, among other environmental woes. This combination worked better on bed bugs than pyrethroids alone because the active ingredients attacked the bugs' nervous system in different ways, although lab tests were already showing resistance in some populations. Other pesticides were reregistered for bed bug use. One was chlorfenapyr, which works by disrupting the way the body uses energy, although it kills more slowly than the nerve-agent pesticides. Another option was a class called growth regulators, which mimic hormones related to development to stop the bugs from reaching adulthood and thus prevent them from reproducing, although the stunted bugs could bite even if they couldn't breed. While this idea has been around since the thirties—discovered, in part, by the aptly named Cambridge entomologist Sir Vincent Brian Wigglesworth—new research suggests that these growth inhibitors only worked on bed bugs at doses far higher than allowed by their labels. And there were other products called synergists that help certain insecticides work better. If new options came to market, however, and each acted on different molecular pathways inside the bugs and involved different genes or proteins or transcripts, there was a better chance to avoid resistance. Maybe, possibly, if only. No one would know for sure until they tried.

My conversation with Beal kept circling back to the people of Ohio, who were, he pointed out, "mixing and matching and pouring things together that they shouldn't be." There were the accidental fire starters and the pesticide abusers, and also people who were using so much diatomaceous earth that the clouds of dust were allegedly getting sucked into their lungs. Beal told me about families spending their food budgets on insecticide. Other people didn't care if their attacks on the bed bugs were even practical, as evidenced by an Ohio man who, after being told by professionals that his constant use of aerosol foggers would do nothing but scatter his bugs throughout his apartment building, responded: "I don't care as long as they get out of *my* space."

"Drastic people will take drastic measures," said Beal. He meant the frequent bed bug accidents, but the sentiment was just as easily applied to propoxur.

• • •

Long before Ohio's mixed and matched pesticides, there were other bed bug remedies, which I began to collect, growing increasingly proud of my cache of historical examples. Some were just as regrettable as our modern tragedies; others were simply odd. The oldest written references I found are the spells cast by the Egyptians to keep the bugs away and the works of the ancient Greeks, who hung hare or stag feet off of their beds in a similar appeal to magic. In more recent centuries, and with less fancy, bed bug victims tried barricades and traps. One homemade version involved planting each bed leg in a saucer filled with paraffin or kerosene, which made it difficult for bed bugs to climb to the mattress. There were also wicker baskets or spruce planks drilled with holes that could be attached to a bed frame to provide attractive harborages for the bugs, which would be smashed to death the following morning. And a Balkan folk remedy documented in a 1927 report from the Imperial and Royal Austro-Hungarian army—which I could only find written up in German and had to manually transcribe and then translate—suggested placing bean leaves under the bed. Tiny hairs on the leaves snared any bed bug that walked through, keeping them prisoner until someone could sweep up the leaves and burn them. This approach may have been much older and widespread. In the late 1670s, the English philosopher John Locke wrote that putting leaves of dried kidney beans under the pillow or around the bed would prevent bed bugs, and in 1777 King Louis XV's secretary wrote a letter to a French newspaper recommending either bean or comfrey leaves for the same purpose.

Not long after the first patent was filed in the United States in 1790, people began making their bed bug traps and bug-proof beds official. By 1920 such inventions were so common that the US Patent Office created a special designation to track them: class 43, subclass 58, subclass 107, subclass 123, or "Fishing, Trapping, and Vermin Destroying: Traps: Insects: Bedbug Type."

More reliable and far more dangerous than traps were poisons, made more common with the professionalization of pest control. In the late seventeenth century in England, some of the first known specialist exterminators set up shop. Most famous was the family company Tiffin and Sons, the self-proclaimed "Bug-destroyers to Her Majesty and the Royal Family" whose London shop sign read: "May the Destroyers of Peace Be Destroyed by Us." The Tiffins, who were in business for at least a hundred years, used secret mixtures to treat only the finest homes, proof that even the rich got bed bugs. In one historical account, one of the Tiffins recalled treating the bed of Princess Charlotte, daughter of King George III. When Tiffin found a bed bug and showed it to her royal highness, the story goes, she said, "Oh, the nasty thing! That's what tormented me last night; don't let him escape." Tiffin reportedly said that the bed bug "looked all the better for having tasted royal blood!"

The Tiffins' toxic bed bug formula may have been similar to other poisons of the era, such as John Southall's Nonpareil Liquor, a recipe that Southall, an English gentleman scientist, claimed

An early patent for a bed bug trap, published in 1865. Credit: United States Patent and Trademark Office.

he got from an old Jamaican man during a trip to the Caribbean. When Southall returned to England, he brought the recipe with him and sold the liquor widely. He also collected bed bugs in glass jars to study them and published the earliest known scientific work on the insects in 1730. The book, *A Treatise of Buggs*, cost just a shilling when it published, and it remained popular for the next 150 years. Reprints are still available today—I bought mine on Amazon.com for fourteen dollars and found another edition at the New York Public Library for free. Inside both books is my favorite, if whimsical, description of a bed bug, which is based on what Southall saw the first time he put the insect under a microscope: "A Bug's body is shaped and shelled, and the shell is transparent and finely striped as the most beautiful amphibious turtle. [It] has six legs most exactly shaped, jointed and bristled as the legs of a crab. Its neck and head much resembles a toad's. On its head are three horns, picqued [*sic*] and bristled, and at the end of their nose they have a sting sharper and much smaller than a bee's."

Southall's original recipe for his liquor is lost, although its secret ingredient may have been the Jamaica quassia, a tropical plant that can be used as an insecticide. A 1793 republication of his treatise included an appendix with a new liquor recipe made from ingredients available in England: herb Robert, wild mint, wood spurge, and fly mushrooms. It may have been the poisonous mushrooms that gave the formula its lethal character. More likely, it was the addition of mercuric chloride, a highly toxic form of the poisonous element mercury.

Mercury wasn't the craziest thing we used to kill bed bugs, though it's hard to pick a winner for that category. For me, a front-runner is the blowtorch. Michigan public housing superintendents blasted their numerous iron beds with these in the early 1900s, and in the thirties, recent Naval Academy graduates on the USS *Arizona* did the same to their bunks. Another fantastic example comes from Ohio and foreshadows the state's current woes. In 1907 fire marshals reported a rash of blazes set by people who, having drenched their mattress with gasoline to kill their bed bugs, struck a match to see if they'd missed a crack or crevice. Boom.

There is also a collection of odd cures in a late eighteenth-century American guide with the unwieldy name *The Complete*

Vermin-Killer: A Valuable and Useful Companion for Families, in Town and Country: Containing safe and quick Methods of Destroying Bugs, Lice, Fleas, Rats, Mice, Moles, Weasels, Caterpillars, Frogs, Pismires, Snails, Flies, Moths, Earwigs, Wasps, Pole-cats, Badgers, Foxes, Otters, and Fish and Birds of All Kinds &c. &c. to Which Are Added Useful Family Receipts for the Preparation of Medicines, for the Cure of Common Disorders. The Gentlemen Farrier; or, Directions for the Purchase, Management and Cure of Horses. The Compendious Gardener and Husbandman; or, Observations Relative to Gardening, Husbandry, &c. With divers other Matters, well worthy the Notice of the Country Gentlemen, the Farmer, and every House-keeper. To kill bed bugs, said the guide, simply spread some gunpowder in the cracks of the bed, then light it on fire. If that doesn't work, the guide suggested placing boiled rabbit guts under the bed or washing the bed frame with wormwood and hellebore, a poisonous flower, that have been boiled in a "proper quantity of Urine."

In the 1800s and early 1900s in the United States, noxious sprays made of arsenic and mercury were common, as were mixtures of mercuric chloride and egg whites or flammable liquids including turpentine, gasoline, and kerosene, which housewives brushed into cracks, crevices, and other bed bug havens with feathers. Safer poisons were also available, such as pyrethrum, the ancient powder made from dried chrysanthemum flowers.

Historically, the best method for extreme infestations was probably fumigation, which pest controllers used in houses, trains, and ships. Many of these gases were just as dangerous as the other poisons. In Great Britain in the 1900s, one popular fumigant was heavy naphtha, a combustive vapor made from distilled coal tar that was typically released in a well-sealed and heated room. Another involved burning a bowl full of powdered sulfur (the "fire and brimstone method"), which marred furniture and fabrics and was also a fire hazard. Still other fumigants were linked to cancer, such as ethylene oxide, which is also highly flammable, or contained deadly hydrogen cyanide such as the infamous Zyklon B that the Nazis used in gas chambers during the Holocaust. Exterminators using this approach were an unsettling specter in their post-apocalyptic respirators. Even these were infested: photos from Camp Lee, Virginia, in 1943 show bed bug detritus in the folds of

A man in a gas mask opens a can of fumigant with a special can opener at Camp Lee, Virginia, 1946. Credit: LTC W. J. Rogers, courtesy of the US Armed Forces Pest Management Board.

a US Army gas mask, and a military flyer from the era reminded recruits to inspect their gear regularly.

Danger aside, there was another problem common to all historical bed bug remedies. The traps only caught insects that walked through the right place at the right time. Flammable petroleum liquids only killed on contact, evaporating before the bugs emerged for their nighttime meals; pyrethrum, too, provided only a short-lived lethality. Even George Washington Carver, one of America's most celebrated inventors, couldn't come up with a fail-safe bed bug poison. His attempt in the early twentieth century worked only for a short while before fading away into uselessness. Fumigation and heat were also ephemeral killers, disappearing once they dissipated into the air. If bed bugs crept back in a room in a bag or laundry basket after even the most successful treatment, it was back to the beginning.

DDT, of course, eventually provided a long-lasting residue, as did pyrethroids and the other synthetic insecticides, but the re-

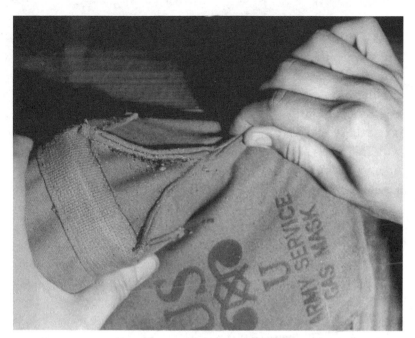

A bed-bug-infested army gas mask at Camp Lee, Virginia, 1946. Credit LTC W. J. Rogers, courtesy of the US Armed Forces Pest Management Board.

lief was only temporary. Resistance or environmental and human health concerns eventually caught up to them all.

• • •

As I collected documentation on old methods of bed bug annihilation and learned about risk cups and modern insecticides, I wondered whether it would really be all that difficult to come up with a new solution. Couldn't the chemical companies simply mix up a concoction better than pyrethroids or propoxur or even DDT? To find out, I flew to Germany to visit the Bayer CropScience headquarters, a subgroup of one of the world's largest chemical companies.

The headquarters are located in Monheim am Rhein, a tiny village on the banks of the Rhine halfway between Dusseldorf and Cologne. I showed up thirty minutes late to the security gate due to a combination of jet lag, bad weather, and a cabdriver who mistakenly thought I wanted to go to a different Bayer complex that was located in the opposite direction from my hotel. When I finally arrived, flustered and damp from the rain, one of my hosts, a product development manager named Volker Gutsmann, wearing a neat blue dress shirt and geometric glasses, was pacing by the front desk. He handed me an ID badge and whisked me onto the company's lush and sprawling campus.

One of our first stops was the chemical library, located in a warehouse that could store a Boeing 737. I walked through the cavernous building with Gutsmann and a research scientist named Martin Adamczewski, who led us through a series of heavy doors and up a metal catwalk. Below us, stretching to the far reaches of the airy space, were tall racks holding several million glass bottles of powdered chemicals. Adamczewski explained that Bayer CropScience had manufactured about half of the samples over the past two or three decades. The rest were purchased from other companies. Just a few milligrams of each chemical, roughly the equivalent of a pinch of salt, sat in each of the small brown glass jars, which stood like soldiers on red removable shelves stacked on the racks. When a Bayer scientist wants to test one of these compounds to see if it might be a good candidate for an insecticide, Adamczewski said, he or she orders it through a custom online form, which wakes up the library's army of robots.

I watched from the catwalk as the robots clanked with life. One, a simple platform, whirred down a track in the alleyway between the chemical racks. Once the platform was in line with one of the racks, a pneumatic lift shot it into the air until it was parallel to the shelf that held the brown bottle containing the chemical it had been programmed to find. A conveyor belt fired and the shelf slid from the rack to the robot, which dropped down, raced back along the track, and delivered the shelf to another series of robots.

The Bayer scientists explained that I wouldn't be able to see all of the next steps, which took place in two separate rooms. The first was out of my view, blocked off from visitors, and the second wasn't operating when I walked by; its robots were silent, frozen in place. Adamczewski explained that the robots, if I could see them in action, would pick up the glass bottles of chemicals with metal grippers and deliver them to the grasping claws of yet more robots, which would twirl the bottles in front of scanners to read a barcode identification sticker, add a solvent, pipette near-invisible drops of

The Bayer CropScience compound library. A robot runs up and down a track (*center*) to retrieve small bottles of powdered compounds stored on racks. Credit: Bayer CropScience.

the dissolved chemical into several hundred tiny wells embedded in flat pieces of plastic called a microtiter plate, and then seal the plate with foil. A few days after a Bayer scientist placed a chemical order, a courier would deliver the microtiter plates to his or her lab.

When a researcher at Bayer is looking for a new insecticide for a specific pest insect, around half of the compounds in the library are available for testing, although it varies for specific projects. (At other companies, such a selection process could also be done through more conventional means including university research or literature searches, like those that Paul Müller did as he combed old scientific papers in his search for DDT.) Next, the chemicals from the library go through a high throughput screening process that is also used to search for promising new compounds by the pharmaceutical industry—Bayer, inventor of aspirin, is also one of the world's largest drugmakers. High throughput screening uses automated machinery and computers to run thousands of biological or chemical tests simultaneously. This is particularly handy when researchers are searching through a large set of chemicals with no prior knowledge of what they look like or how they work, as is common for insecticide discovery.

One high-throughput setup at Bayer can test between 100 and 150 microtiter plates every day, each of which contains 384 tiny wells, several hundred of which may be filled depending on the experiment. If all the machines were running at once, which isn't typical, around 200,000 chemicals could be tested in a single week. Less than one hundred years ago, Müller toiled for four years to screen just a few hundred molecules on his blue bottle flies.

We left the chemical library to continue on my tour, dodging raindrops across a grass courtyard to another building. There, in a narrow room, the Bayer scientists explained to me the first round of high-throughput insecticide screening. I watched another set of robots, which were basically articulated arms on a sliding track, as they took the microtiter plates filled with the chemicals from the library, opened them, mixed their contents with a target material, and then monitored and recorded the results. In this particular test, the target was a single enzyme that is found inside an insect, and a potentially successful molecule will bind to it and cause it to malfunction.

In a larger room down the hall, I saw how a more typical insecticide test would take place. The robots in the room were virtually the same, but here, if the robots had been in action that day, each of the tiny wells in the microtiter plate would have held living cells that had been genetically engineered to contain ion channels just like those in an insect's nervous system. The robot would add the chemicals from the library to these constructed ion channels while simultaneously monitoring their changes with sensitive cameras that capture the flash of an electrical response, represented by a dye that responds to the change in charge. This would show whether or not the ion channels could function in the presence of the chemical. The scientists would later measure the strength of each response with custom image analysis software that measures the depth of the color of the dye. The chemicals that allowed the ion channels to function normally would be dismissed from the pool of insecticide candidates.

Between five hundred and ten thousand chemicals make it through the initial round of tests, which, even with the help of high-throughput screening, takes between six weeks to four months. The scientists reorder this subset from the whirring and clanking robots in the molecule library and then put the chemicals through another series of experiments. These are dose response tests, and they explore how the original target, whether it was an enzyme or an ion channel, responds to different concentrations of each chemical and ranks them by potency. Just a few hundred chemicals make it to the next round, where they are scrutinized to see if they react only with their intended target or to many targets. In the latter case, for example, the molecule may never make it to a real insect's nervous system because it would react with other proteins first, whether in the insect's exoskeleton or somewhere else in the body. These, too, are dismissed.

All of the tests up to this point are akin to a recruitment process, where the obvious bad matches are weeded out. The real experiments, the Bayer scientists explained, haven't even started yet. And anywhere between zero and several hundred chemicals make it to this stage; still, these are nothing more than compounds with a whiff of possibility, requiring another series of complicated tests.

In yet another building, I sat down with a third Bayer scientist,

Sebastian Horstmann, a tall slender man who spoke so softly that I had to sit very still to hear him. Horstmann stepped me through a presentation on his computer explaining the next cascade of experiments. After ordering more test chemicals from the library robots, a new set of scientists run them through more high-throughput robots, which measure how well the chemicals kill whole insect larvae or damage other insect body parts. Some aren't up for the task and are dismissed. Still more are removed because more tests show they work along the same molecular pathway as current pesticides, including pyrethroids; because so many pest insects are likely already resistant to these, it's not worth taking them any further. More are thrown out because they kill too slowly, which means they aren't good enough for a product.

When we made it to the end of the slides, Horstmann stood up from the desk and led me down a long hallway into a series of rooms where the best chemicals are tested on adult insects. These insects are raised from birth in a high-tech insectary at the opposite end of the long hallway—essentially an insect nursery, with several rooms filled with bins of sleek and well-fed cockroaches, ants, and bed bugs. Some of the bed bugs were susceptible to insecticides, and I felt a shiver of awe when I peeked into their jars and realized I was looking at Harold Harlan's bed bugs' German counterparts, which had been kept at Bayer for decades, leftover from a previous era of insecticide testing. In the testing rooms, there were cabinets filled will various materials you might find in a house, including tiles of various materials, pieces of varnished and natural wood, and wallpaper swaths. There were also various experiments set up, with containers of mosquitoes sitting on top of tiles that had been painted with different concentrations of a possible new insecticide.

Horstmann told me that while the live insect experiments are under way, there are simultaneous safety studies. The chemicals are fed to rats and squirted in cell cultures to see if either dies and at what point. Micronucleus tests show if the chemicals hinder the organization of chromosomes during cell division, which would hint at possible damage to a growing embryo. There are tests to show the chemical's ability to cause cancer or to mimic a hormone in the human body, which might interfere with reproductive health and development. Still more look at the impact on the sodium channels found in the human nervous system, grown outside the body in a lab, to see if they will grip open the gates and cause people to tremor to death. A similar barrage of tests for ecological toxicology explores how the chemical might impact the environment.

Only a few chemicals make it through to the next stages; sometimes there is only one lonely candidate left, or even none at all. If there are any finalists, they are formulated with other ingredients to see how they might be packaged in a product. Horstmann explained that his team also studies these chemicals in field tests to see how well they work in the real world, where an insect's behavior isn't limited to the controlled setting of the lab. Next, if any of the compounds are still working as planned, Bayer sends samples to global partners at other companies and universities to see how it fares in different countries and climates. If and when it passes these trials, and only then, the company applies for registration with the EPA in the United States and its counterparts worldwide.

Once these regulatory groups have scrutinized the product, the toxicology data, and the efficacy data, they either reject it or give it a label. At this point, nearly ten years and more than $256 million later, the product is allowed on store shelves, ready to do its job in the real world. Now the company must race to make back its investment before its patents expire, when generic companies can skip the expensive insecticide discovery and copy the formula's blueprint.

Bayer isn't alone in the complicated business of insecticide discovery. The process happens, more or less, in ten to fifteen major companies worldwide, including BASF, Dow, DuPont, and Syngenta. I talked to scientists from several of these places about the likelihood of finding an insecticide that could kill bed bugs. Even once the companies discover a promising chemical, the scientists said, they don't dive into insecticide development unless they can reasonably predict a good return on their investment, which means figuring out which insects the chemical can kill and how much money they can make off of each category. There are essentially three tiers: agricultural pests, disease vectors, and a miscellany of less threatening, but bothersome, insects. Around 80 to 90 percent of insecticides ultimately target agricultural pests, which command the biggest market and thus get the most industrial attention.

If a chemical can't kill any insects that threaten agriculture, it has little chance of becoming a product. If a chemical is active against at least some major agricultural insects, then the company looks at the next tier to see if it will also work against malaria mosquitoes or other insects that spread disease. And if a chemical is found effective in these more lucrative tiers, the company looks at the third tier. This is where the lowly bed bug sits.

From a business perspective, the logic is reasonable. The physical real estate that bed bugs occupy is tiny compared to the vast expanses of farmland and orchards—think of how much space all the houses and apartments in the world take up, and then compare it to the wide-open fields that grow our food. The amount of pesticide that can be used in our small domestic space, and especially the bedroom, is minuscule. Bed bugs are not known to spread disease, so killing them isn't considered a major worldwide public health goal. As far as the chemical companies are concerned, the bed bug is not what you'd call a moneymaking pest.

In addition to the unlikely case that a brand-new insecticide may target bed bugs, the scientists told me that companies are testing existing products to see if these might kill the pests. If the companies find anything promising, they can apply to the EPA for a new label that includes bed bugs or bedrooms. This is cheaper than finding a new insecticide because the companies can piggyback on the risk studies for agricultural and vector use, but it still may cost up to half a million dollars. The price tag covers the development of a new indoor-use formula appropriate for bed bugs and the fees to provide data to the EPA proving that it works. To make up for this investment, as well as for the money put into manufacturing the product and putting it on shelves, new EPA-registered bed bug products will cost more than the pyrethroids that have already been available for decades. This will likely be the case until the production facilities mature. Higher production means higher volume, which is part of what drives an insecticide's price. (Another factor is how much the consumer is willing to pay.) Only then, years later, consumers might enjoy a relatively new and cheap pesticide that can help wipe out bed bug infestations. Once a product becomes popular, of course, it has an increasingly shortened shelf life thanks to the threat of resistance. Two decades ago, there were more than five hundred known species of pesticide-resistant insects and mites. Today the estimate has risen to six hundred at a cost of $60 billion per year.

Back in New York, discouraged about the prospects of a new bed bug insecticide, I began to read about alternatives. I learned that some companies enjoy a loophole by choosing their active chemicals from among thirty-one "minimum risk" ingredients, which are excluded from the EPA's expensive toxicity requirements and therefore bypass hundreds of millions of investment dollars. These are the 25b ingredients, named for the section of the Federal Insecticide, Fungicide, and Rodenticide Act under which they are listed, and they include essential oils such as cedar, citronella, and peppermint; mild acids from apples, oranges, and lemons; and other relatively benign ingredients from table salt to dried blood to rotting eggs.

The EPA deemed 25b ingredients demonstratively safe in 1988 to make it easier for companies to bring related products to market. The conditions for making and selling 25b products require the active ingredients, or those that allegedly kill insects, to be on the pre-approved list and the non-active ingredients to be on a similar list. The product label must list all of these ingredients as well as the concentration of the active ones. Companies also aren't allowed to make public health claims that link a specific pest with a disease or to tell lies on their product labels.

I've seen 25b products on the shelves of hardware stores and my local pharmacy. Some are sold in box home goods stores, which strategically stock the bottles next to bed bug mattress covers. The products generally boast slogans like *All Natural! Kills Bed Bugs on Contact! EPA Approved!* While technically true, I found that the claims are misleading. It's true that the products' active ingredients are natural as opposed to synthetic, although natural doesn't necessarily mean a product is safe or good and is instead often used as a marketing ploy. It's also probably true that the products kill on contact, but without a residual layer that sticks to a surface, this only helps if you are ready with the spray at the precise moment the bed bugs reveal themselves at night. ("At that point," an exasperated entomologist once told me, "you might as well save your money and hit the thing with a hammer.") And while it is also true that the 25b products are EPA Approved for safety, the catch is that the EPA doesn't require any efficacy data. In other words, the EPA makes no claims that the 25b products work.

• • •

If the chemicals can't save us, what will? Perhaps there could be another way to go about killing off bed bugs by giving an old technique a new twist. Researchers are still trying to build a better bed bug trap, just as inventors and tinkerers did hundreds of years ago. I pored through current research to find out what they were up to. At the entomology conference I went to in Knoxville, where I saw the presentation about Usinger's bed bugs and population genetics, another talk caught my eye. An insect biomechanics researcher at the University of California, Irvine, had run bed bug after bed bug across the underside of kidney bean leaves to understand how John Locke and people in the Balkans were able to trap bed bugs with these plants. In the forties, scientists from the US Department of Agriculture showed that the small hairs on the leaves, called trichomes, might work like the hooked side of a Velcro strip by grappling the stiff hair on bed bugs' legs. The California scientists proved this wrong by looking at bugs stuck to bean leaves under powerful scanning electron microscopes. At the conference, I was transfixed by an image that showed the sharp trichomes actually piercing a bed bug's foot like meat hooks, tortuously immobilizing it. A synthetic version, the scientist suggested, could be manufactured and sold as a trap.

Meanwhile, I found research by scientists at Stony Brook University in New York that involved weaving webs of synthetic strands thinner than human hairs to try to ensnare the bugs. And several competing companies are experimenting with a modern plastic version of the saucers that used to be placed under bed legs, some of which come in different colors (whether bed bugs are more attracted to red or to black, I learned, is a point of debate).

There are also scientists playing with lures designed to mimic bed bug pheromones to trick the insects into thinking their broth-

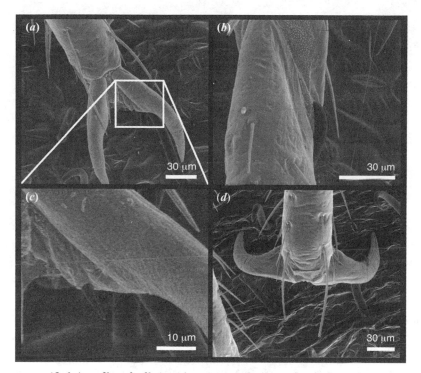

A magnified view of bean leaf hairs either piercing a bed bug's foot (*a*, *b*, and *c*) or hooking around them (*d*). Magnification via a low-vacuum scanning electron microscope. Credit: University of California, Irvine.

ers and sisters are sending them signals to act in a specific way. Scientists at the London School of Hygiene and Tropical Medicine, for example, have collected sample after sample of bed bug poop, convinced that the feces contain the aggregating pheromones that help guide the insects home after a blood meal. As the small odorous molecules emanating from the poop waft through the air, they attach to chemical sensors on the bugs' antennae. The bugs walk toward the increasingly intense smell until they reach their refuge, where the concentrations are greatest. The scientists are working to isolate these compounds to add to a sticky trap, which will bring the bugs in and then keep them there.

The trouble is, modern bed bug traps face the same problem as all those that came before: in order to work, bed bugs must walk through them. Even when they do, it proves the existence of the bugs in a room, but it doesn't mean it has trapped or killed them all. This requires an inescapable mode of death. Since synthetic chemical killers are expensive and elusive, some scientists have looked for other ways to poison the insects.

Scientists from Penn State University think the answer may lie in the fungus *Beauveria bassiana*, which lives in soil throughout the world. *B. bassiana* is a biopesticide—a natural insecticide found in living things including animals, bacteria, and plants, or from minerals or oils. The fungus's fine white dusty spores have a natural affinity for fatty surfaces. When certain insects come into contact with the spores, either through contaminated soil or other infected insects, the spores stick to the outermost layer of the waxy exoskeleton. The moisture on the insect's body allows the spores to sprout, and they penetrate into the exoskeleton. From there, the fungus spreads, blooming in the insect's circulatory system and clogging it.

In the Penn State tests, the scientists sprayed various surfaces, including smooth printer paper and textured jersey knit, with a milky mixture of the fungus and dropped bed bugs on top. Both Harlan bugs and a strain collected from the field died within three to six days. But to make it as a product, the fungus would have to pass EPA requirements for biopesticides, which are just as stringent as those for synthetic pesticides. Here, the agency is concerned with not only toxicity, but also hypersensitivity and infectivity. This means the agency wants proof that the fungus won't irritate a person's skin or pierce it and bloom inside the blood.

I also found research attempting to attack bed bugs in a place they will undoubtedly always return: blood, the same sustenance that has driven their spread for so many millennia. In 2012 I interviewed a Virginia emergency room doctor who had dosed himself and a few volunteers with ivermectin, an anti-parasitic drug common to heartworm preventatives for dogs and livestock. The drug has also been used since the late eighties to treat humans for river blindness, a disease common in sub-Saharan Africa caused by different species of roundworms, and is sometimes prescribed to treat head lice, pubic lice, and scabies.

After taking the drug, the doctor and his test subjects fed live bed bugs on their arms. The ivermectin killed some of the bugs but failed to get them all, and in real life far longer treatments would be necessary than is currently proven safe. In other words: do not try this at home. (Merck, the maker of the drug that the doctor tested, politely declined to comment when I sought the company's take on the experiment, and it is doubtful whether it will pursue clinical tests. New labels for drugs are even more expensive and time-consuming than for pesticides.)

If blood can't be easily or safely tainted, another option is to make it useless once it hits the bed bug's digestive track. In 2009 scientists at the National Institute of Advanced Industrial Science and Technology in Tsukuba, Japan, showed that damaging symbiotic bacteria that live in the bed bug's gut is bad for its health. The bacteria come from the genus *Wolbachia*, named for one of the scientists who discovered it in the bowels of a species of mosquito in 1924. *Wolbachia* are among the most common beneficial bacteria that live in insects and are found in around two-thirds of the known insect species on the planet. The bacteria help pull key nutrients from food or aid in reproduction. *Wolbachia* are one example out of an entire microbial ecosystem that flourishes in insects, including *Wigglesworthia glossinidia*, which helps the tsetse fly synthesize vitamin B and is named after Sir Wigglesworth of bug hormone fame. Humans, too, have gut microbes that help produce vitamins key to our health.

The Japanese scientists suspected that *Wolbachia* similarly allows bed bugs to synthesize vitamin B. To test the idea, they fed rabbit blood laced with antibiotics to bed bugs in the lab, which wiped out the *Wolbachia* inside. Bed bugs without the symbiotic bacteria were stunted and sterile; both traits were reversed after the scientists fed the bugs blood supplemented with vitamin B. Whether the drugs could kill bed bugs in a real-life setting is unclear. Dosing a person with antibiotics could add to the growing problem of drug-resistant bacteria, adding yet another negative layer to the bed bug predicament. The bed bugs' *Wolbachia* may grow resistant, too.

All of the basic research is interesting, but none could get rid of a bed bug infestation completely on its own—instead, it would need to be incorporated into existing integrated control tactics.

But even so, the best traps, chemicals, or biological manipulations will never overcome an obstacle that Dini Miller pointed out to me when I visited her in Virginia.

"The hardest thing to control isn't the bed bugs," she said.

"What is?" I asked.

"It's the people."

Indeed, people hide bed bugs from landlords, skip treatments, bypass instructions on insecticide labels, fail to heed their pest controller's guidance, and ignore sage advice on avoiding the bugs to begin with. Others accidentally set their mattress on fire with a bottle of rubbing alcohol and a match. And no technology can save us from that.

SIX

FEAR
When Things Go Bite in the Night

In the corner of the freezer in a Brooklyn apartment, next to a nearly empty three-year-old carton of ice cream, Anoki had stashed two bed bugs. They were in a plastic takeout container triple-wrapped in cheap plastic sandwich bags, and they had been dead for a long time, having sat there for more than a year. Anoki (not his real name because I promised not to use it) originally put the bugs in the freezer in case his landlady needed proof before she would pay his extermination bill. But then he left them there. He did this for the same reason that he had kept the last bite of the LaSalle Vanilla Swiss Almond, leftover from the first time he'd ever shared ice cream with his girlfriend, Daelyn (also not her real name) on one of their early dates: it was the physical representation of a significant memory. The ice cream was a happy one; the bed bugs were not.

"He didn't even let me know that he put the bed bugs in the freezer," Daelyn told me when I met the couple on a snowy January day more than a year after they'd gone through their bed bug infestation. "And I was like, 'Why is this empty bag in here?' And then I opened it and I was like, 'Oh, okay,'" she said, laughing.

"And you put it back?" I asked.

"Yeah. I knew why he was keeping them."

I met Anoki and Daelyn through a mutual friend when I was going through a phase of wanting to talk to as many bed bug survivors as possible. We arranged to meet at a coffee shop a few blocks from their apartment, but when we arrived it was filled with silent people tapping on laptops and curled over steaming cups—too quiet for a candid conversation about infestations and bloodsuckers. We moved to a noisy bar next door and ordered a round of hot toddies and stouts. Anoki and Daelyn, both young and hip in knit caps and plastic-rimmed glasses, told me their story.

Anoki first started getting bites around a year and three months earlier, not long after he and Daelyn moved in together. At first he thought the red bumps on his stomach were just a series of rashes, but the itch was fierce and they kept coming back in different places. His doctor gave him antihistamine pills, creams, and a referral for a dermatologist, who gave him more cream and pushed for steroid shots for the swelling. But neither doctor could tell Anoki the cause of the increasingly frequent bumps or how to stop them.

Daelyn had no bites and was busy with her first semester at graduate school, and so Anoki was left alone to scour the Internet for clues. His searches led to websites talking about bed bugs with pictures of bites. He ripped off the sheets and mattress cover to inspect the bed but found nothing. Weeks went by. Each morning brought a new outbreak of bumps, which spread from his stomach to his arms and shoulders. He called the landlady, and she brought in a father-son pest control team. Not long after the son of the duo arrived for an inspection, he lifted the mattress and exposed four bed bugs sitting in a spot that Anoki swore he had already checked. Two of these went in the freezer; the other two were squished and thrown away.

Then came a whirlwind of stressful cleaning and purging, made worse by the stench of a pesticide-drenched apartment, an upcoming visit from Daelyn's family, Daelyn's finals, living with a significant other for the first time, an already-tight budget stretched too thin, and no money to buy a new bed. Each piece of lint sparked a frenzy of new inspections; each torn jumbo trash bag spiraled into worrying whether the clothes inside were now contaminated, or whether they could avoid another trudge to the Laundromat.

Daelyn's proclivity toward anxiety and depression made her guilt over the bugs especially weighty. Had they come with her during the move, she wondered, or from a used chair they had bought together? Anoki had his own struggles, including obsessive-compulsive disorder. Still, his therapist commented on how well he was handling the bed bugs; this was especially evident when Anoki, after the bed bugs appeared to be gone, let the once-incessant inspections and cleanings fade. He had no new bites. The apartment was back in shape. It took mental acrobatics, but he knew it was time to move on.

And so Anoki found himself staring at the bed bugs in his freezer just a few weeks before our meeting. He and Daclyn had finally been able to afford a new bed and had just thrown out their old mattress after wrapping it five times in plastic painter's sheets secured with heavy-duty tape, despite the fact it had been deemed bed-bug-free for a long time. He took the takeout container that held the bugs, left the apartment, and walked several blocks before dropping them into a sidewalk trash can.

"How did it feel to finally get rid of them?" I asked.

"It felt amazing," he said.

As I listened to the couple's story, I thought about all of the other bed bug rituals and compulsions people had told me over the previous months. Hiding bed bugs in the freezer, in context, isn't so strange. There is the apartment building owner who downs antihistamines like a nightcap before bed, which he told me calms his nerves better than any alcohol can; the family who got rid of nearly everything they owned after a bed bug scare, including both of their cars; and the journalist who threw out all of her newspaper clippings, which spanned many years of work, and dozens of personal journals, even though her exterminator said she never had bed bugs to begin with. The first time I met him, a friend of a friend revealed to me that his wife, who has a deep fear of uncleanliness, grew so worried about the stories of bed bugs in theaters that she eventually refused to go to the movies unless they both wore protective polyethylene suits—the kind people use when they're doing lead or asbestos decontamination. The couple tried it once, much to the unease of fellow theatergoers, and then stopped going to the movies entirely. Another friend who lives down the street from me had bed bugs, which her roommate's boyfriend apparently brought in from his own infested apartment, and for months after a successful treatment, she would text photos of spider beetles and baby cockroaches to her exterminator along with the message: "Bed bug??????"

Then there is the woman who got bitten by bed bugs in a New Orleans hotel room during a conference. She was worried she had brought some of the insects back to her Washington, DC, apartment, which she shared with four roommates. To prove to herself that none of her dozens of bites were new, and therefore possibly from a bug that had slipped home in her suitcase, she took a pen and started a complicated tracking system. First she made a small circle next to each bite in blue ink. The next day, she drew a triangle next to all the bites, noting those that had no blue circle. The next day, it was tiny squares. By the end of the week her body was so covered in ink that it wouldn't come off in the shower.

There is also one of my favorite stories from a friend who had bed bugs in the New York apartment she shared with her boyfriend. The whole building was infested; the landlord was no help; the bugs required three treatments—things were tense. During an argument amongst the bagged clothes and the white drifts of diatomaceous earth, the boyfriend locked himself in the bathroom. An hour later, at which point my friend had almost given up and left him to his apparent tantrum, he emerged wearing nothing but his boxer shorts and said, "Maybe this will help us see the bites." He hadn't been hiding out of anger. Instead, frustrated with the bed bugs, he had been busily shaving off all of his body hair.

I empathize with these stories. During one of my bouts with bed bugs, I remember lying awake at night covered in insect repellent and staring at my ceiling, twitching at the slightest sensation on my leg or my arm. I had pulled the bed away from the wall, stripped off all of the blankets and sheets, and created a barrier made from double-sided tape on the floor. And I still got bitten. Another time, I forced a roommate to help me throw out my favorite upholstered chair because I was convinced bed bugs were biting me when I took naps on it. And I've also spent hours meticulously vacuuming each of the several hundred books in my library and steaming the internal seams of my dresser, inch by inch.

That an insect would inspire unease is no surprise. An estimated 19 million people suffer from an irrational fear of insects, or entomophobia. While many more don't suffer from the same paralyzing anxiety, they still feel revulsion or disgust when they see an insect. According to the philosopher and insect ecologist Jeffrey Lockwood in his book *The Infested Mind*, these emotional responses stem from two places: our evolutionary past and our cultural present. Evolutionary theory suggests that we are primed to be frightened by the quick skittering movements and alien forms of insects. Some, after all, are indeed dangerous and can deliver a venomous bite or transmit a deadly disease. It may have made more sense for our ancestors to avoid all insects—and even things that move or feel like them—than to be indifferent and thus vulnerable. The hominid that jumps at a harmless butterfly may feel silly but will have a better chance at living on to have children who are also easily startled. The hominid that reaches into a crevice without looking may die from an unseen sting. Our modern culture, argues Lockwood, nourishes these tendencies. We learn from our families and even our peers how to react to insects. If our parents squealed and grimaced over the sight of a cockroach when we were young, we are more likely to do the same.

Lockwood also notes that our increasingly urban spaces make it easy to avoid insects, or at least the majority of the 10 million species thought to exist worldwide, and this has diluted our ability to separate the dangerous insects from the benign ones. In the modern antiseptic living space, increasingly barricaded from the outside world, the presence of any insect seems like an invasion. And we have entered into next-level expectations for what constitutes as clean, with our antibacterial soaps and arsenals of cleaning sprays. This feeds into a heightened intolerance of even the occasional creepy crawlies.

But people who live near mosquitoes that carry malaria or West Nile virus don't shave their bodies or mark them in ink or cover them in protective Tyvek suits. They also don't dispose of all their worldly belongings or spend hours vacuuming them. Neither do people with cockroaches, rats, or silverfish. Bed bugs inspire behavior beyond what these disease spreaders or common household pests can muster. ("Logic goes out the window," as Anoki told me when we discussed the freezer incident.) What makes bed bugs different?

One possibility is because bed bugs attack in the night. When we sleep, we are vulnerable. Humans rely on a stage of rest called rapid eye movement, which may be important for memory consolidation, learning, and the regulation of emotions. It is also the period of sleep when we have our most intense dreams. During REM,

which happens several times a night, the body is basically paralyzed so we don't physically act out those dreams, which leaves us
defenseless. Evolutionarily speaking, humans and other animals
that experience REM may be primed to seek shelter when they
sleep; once they find it, they clear out debris and, possibly, parasites. They also bring in leaves or other materials that add comfort
or conceal them from a predator. Animals continue to act out these
rituals in modern domestic spaces that are generally safe—a dog
circles its bed, for example, or a person may check that the doors
are locked before turning in for the night. That same urge likely
led our predecessors to caves, at least occasionally. Another way to
look at the origin of bed bugs is that they arose directly from our
need to seek a safe place to sleep.

The significance of our vulnerability during sleep is heightened
by the darkness of night. This has long been lodged in the collective human psyche. Roger Ekirch, a sleep historian, has written:
"Night was man's first necessary evil, our oldest and most haunt-
ing terror." In pre-industrial Western culture, Ekirch writes, the
night provided a cloak of anonymity for murder and other violent acts, as well as for thefts and predatory beasts. It was the time
when unattended fires—the only source of light—might cause a
home to go up in flames. In our past, dark supernatural beings,
too, seemed more active at night: ghosts, witches, and demons.
Families protected themselves from both these real and perceived
threats by shuttering their houses, locking their city gates, and
saying a nightly prayer. To ensure a comfortable sleep, they would
also go on nightly "bug hunts," examining their beds for what
Ekirch calls the unholy trinity of early modern entomology: lice,
fleas, and bed bugs. A clean and safe bed was a sanctuary for sleep
as well as for intimate nighttime conversations and sex.

Although some of our nighttime terrors and rituals have ebbed,
thanks in part to the electric lighting that keeps our world visible
even when the sun goes down, there are residual fears. It doesn't
take much to stoke them. In modern day, the writer Teju Cole beau-
tifully connects this vulnerability to bed bugs in his 2011 novel
Open City, when the protagonist sorts out his feelings on the bugs
thriving in all five boroughs of New York: "The concerns were pri-
meval: the magical power of blood, the hours given over to dreams,

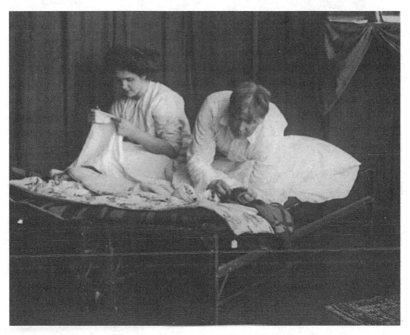

"Dear Friends: We were up most all night looking over things." A couple engaged in a
bug hunt, circa 1907. Credit: Unknown, courtesy of the Library of Congress.

the sanctity of the home, cannibalism, the fear of being attacked
by the unseen." This thought is echoed by nearly everyone I have
asked: *Why are you afraid of bed bugs?* Their answer is a variation on
the fact that the bedroom is a place of repose, a protected sphere
that provides comfort even on the hardest days. It is where we feel
safest to sleep, to close our eyes and let down our defenses. A thing
that not only breaches this sheltered slumber but takes some blood
while it's there feels like more of a violation or threat than the West
Nile mosquito buzzing on the porch or the feverish rat digging
through the trash out back. Bed bugs break our modern illusion of
the cloistered bed.

We also have new expectations for sleep. Just as with our unprecedented new heights of cleanliness, the modern era has ushered
in a different perspective on how and when we rest. We are in bed
for shorter periods of time than our ancestors were, says Ekirch,
and we often think of sleep as a nuisance or a luxury. For the sleep
that we do get, there is more pressure to make it count. Unlike real
threats, such as a deadly predator or an overturned open-flame

"Summer Amusement, Bugg Hunting," 1782. Credit: Thomas Rowlandson, courtesy of the Lewis Walpole Library, Yale University.

lantern, or superstitious ones, such as a witch, a small annoyance like a barking dog or a neighbor's loud television are enough for many of us to declare our sleep ruined. Our pre-industrial ancestors may have had nightly bug hunts, but the bed bugs were a lesser threat to a good sleep; today the bugs have risen as a far more prominent peril.

· · ·

Out of all of the insects, those that feed on blood may strike the deepest fear, possibly since our earliest interactions. According to Michael Lehane's *The Biology of Blood-Sucking in Insects*, bloodsuckers' ability to feed on blood—a rich, high-protein meal—may have evolved at least six unique times during the Jurassic and Cretaceous periods, which together cover 200 to 65 million years ago, when dinosaurs were still alive.

While the origin of the bloodsucking insects isn't known for certain, Lehane suggests two main possibilities. The first involves insects that lived in animal nests long ago. These insects may have accidentally ingested the dead skin and hair of the nest makers while taking meals on whatever they normally ate—maybe fungus or scat. Eventually, perhaps, some of the insects evolved to survive on live skin and then, much later, on blood, which they may have found through open wounds or broken scabs. The second hypothesis points to insects that had already developed sucking mouthparts to feed on plants or other insects; some may have accidentally bitten an early mammal or a reptile and had the right enzymes to digest its blood, birthing offspring capable of doing the same.

Whether blood feeders arose from either of these evolutionary paths or another one entirely, they eventually flourished. When the earliest modern humans arrived around 200,000 years ago, some of these insect species may have shifted focus to this new host. Or, more likely, the insects were already closely associated with our even older relatives and evolved with them. The bed bug's emergence in its hypothetical ancestral bat caves a couple hundred thousand years ago, then, makes it a relative youngster in the larger scale of insects that specialize on blood. Today it is one of around 300 or 400 insect species that regularly feed on human blood, out of 14,000 that feed on blood at all.

For much of our more recent shared history, our squeamishness toward vampiric insect beasts has been well documented in literature and scientific works. Part of that fear was well founded. We long suspected certain bloodsucking insect species of spreading human illness, a hunch that was confirmed in the nineteenth century when science first connected a specific insect to a disease. This proven vector, the mosquito *Culex pipiens quinquefasciatus*, was discovered in 1877 by a Scottish doctor named Patrick Manson, who, while doing research in China, found larvae of a pathogenic filarial worm thriving inside the insect. Following Manson's revelation, scientists would uncover many more pathogenic microbes—including viruses, bacteria, and protozoa—lurking in dozens of other blood-feeding insect species.

These pathogens exploit an insect's body in different ways. Some use it as a vessel to make an exponentially greater number of identical copies of themselves, as with certain viruses and bacteria; some use it to both multiply and morph into a new form, as with malarial protozoans; and others simply change form but don't multiply, as with the filarial larvae that Manson found in his *Culex* mosquitoes, which would eventually grow to be long stringy worms. No matter the style of development and propagation, each pathogen spreads from host to host during bites from an infected insect. They eventually multiply or grow to adulthood in their new host and then burst forth many tiny offspring to be taken up by another insect bite, starting the cycle anew.

In the decades that followed the confirmation of insects as disease vectors, bed bug fear blossomed. Indeed, William Colby Rucker, assistant surgeon general for the US Public Health Service, would proclaim in 1912: ". . . bed bugs are not a disgrace. They are, however a positive danger. Their presence in the house may mean merely accidental introduction. Their continuance in the house means a disregard for health."

To try to prove whether the bed bug was dangerous, scientists poked and prodded the insect. They attempted to infect bed bugs with microbes in order to track whether those microbes could survive and, if so, whether they could jump from one person to another. One example comes from 1904, when a researcher named Herbert Durham from the London School of Tropical Medicine tried to prove whether or not bed bugs were responsible for the spread of beriberi in what is now known as Malaysia. (This loose collection of symptoms is due to the absence of thiamine, or vitamin B1, and may cause everything from difficulty breathing to congestive heart failure to death, but at the time no one knew the cause.) Durham collected bed bugs from the beriberi wards of the Lunatic Asylum and General Hospital in Kuala Lumpur and, every few days, let the bugs feed on a pair of monkeys. The experiment failed to infect the animals. Around the same time, another British scientist stationed in the Malay islands named J. Tertius Clarke took the idea a step further. He crushed about twenty bed bugs into water and injected the mixture under a monkey's skin. Again, at least for the three weeks that Clarke monitored the monkey, "it was still quite well."

From the turn of the twentieth century to around the 1960s, other scientists similarly tried to infect bed bugs with all types of microbes and argued about their findings in the pages of scientific journals. The tests included bacterial infections such as leprosy, typhus, tuberculosis, typhoid fever, and plague; viral infections such as influenza, encephalomyelitis, yellow fever, and polio; and protozoan infections such as Chagas disease, kala-azar, malaria, oriental sore, and sleeping sickness. There were other claims, too, including the idea that the bugs might carry the parasitic filarial worms that spread elephantiasis, which causes colossal swelling of the legs and genitals, and the suggestion that bed bugs were somehow linked to cancer in mice. And in Usinger's monograph on the bugs, he lists thirty diseases that the bed bug was confirmed to carry, but determined: "Cimicidae have been suspected in the transmission of many diseases or disease organisms of man . . . but in most cases, conclusive evidence is lacking."

Even in more recent decades, when bed bugs were rare in parts of the world, the pests were thought to spread newly discovered infections. In the seventies, it was hepatitis B, a virus that had only been discovered in the previous decade and that decimates the liver and is spread, in some cases, by sexual contact. Occasionally, a few bed bugs plucked from larger infestations—including studies done on tropical bed bugs found in an Ivory Coast brothel—would test positive for traces of hepatitis B, although experiments showed that it could not reproduce or survive. In the eighties, the focus shifted to the human immunodeficiency virus, which American and French scientists had linked to the new and frightening acquired immunodeficiency syndrome in 1983. By the late eighties, at the height of the AIDS panic, a virology team from the University of the Witwatersrand in South Africa attempted to give common bed bugs and tropical bed bugs HIV by feeding the insects infected blood. The virus survived between four hours and eight days inside the bugs but did not replicate. Researchers from the Centers for Disease Control's Division of Vector-Borne Viral Diseases in Fort Collins, Colorado, conducted a similar experiment with the tropical bed bug, *Cimex hemipterus*. Both groups concluded that the insects were highly unlikely to spread HIV.

None of the other experiments over the past century successfully linked bed bugs to illness, either; to date, there are no confirmed cases of a bed bug infecting a person with any pestilent microbe. Yet as the bed bug surged over the past decade, the revival of anxiety over its possible role as a vector soon followed. Scientists in North Carolina, for example, explored whether or not bed bugs were responsible for outbreaks of *Bartonella*, the bacteria responsible for cat scratch fever, because the pathogen was emerging in homeless shelters, hostels, and other bed bugs hotspots throughout the southern United States. By 2013 the scientists had found no trace of the bacteria on any of the bed bugs they tested. And an entomologist from the Ohio State University, curious whether bed bugs could spread pathogens in their feces, has spent hours watching the bugs defecate after each meal and noting the time and distance between the two events to calculate how often they leave droppings on their hosts. Perhaps, if it happened routinely and if a pathogen survived in the feces, disease might spread this way.

In 2011 at my first bed bug conference in Chicago, Illinois, I heard about the most controversial of this new line of research, which published as a letter to the editor of *Emerging Infectious Diseases*. In it, Canadian clinicians suggested that the bed bug's comeback may be responsible for the spread of two dangerous bacteria that have evolved immunity to certain antibiotics: methicillin-resistant *Staphylococcus aureus* (MRSA) and vancomycin-resistant *Enterococcus faecium* (VRE). Both resistant bacterial strains arose across the world starting in the eighties and continue to pose a health threat today, particularly in hospitals where they endanger patients with vulnerable immune systems. Bed bugs, the Canadian doctors wrote, came to their hospital with residents from Vancouver's Downtown Eastside, a neighborhood "with high rates of homelessness, poverty, HIV/AIDS, and injection drug use" as well as high rates of MRSA and VRE infections. The doctors collected five bed bugs from three patients from this neighborhood and found MRSA on three bed bugs and VRE on two. But as in the past, the team did not prove that the bed bugs were able to transmit the bacteria from one person to another. At the conference, everyone I spoke with was skeptical of the work, and it still hasn't been confirmed.

It's reasonable to ask whether bed bugs carry disease. Out of the several hundred species of blood-feeding insects that regularly bite humans, it's hard to put a precise number on how many are capable of transmitting contagions and how many illnesses they are collectively able to spread. But the most important forty-odd human diseases spread by insect vectors are associated with dozens of species within around half a dozen major groups, and each sticks to a narrow range of hosts because the organisms have closely evolved and adapted over time. Included in this list are mosquitoes with their malaria, West Nile virus, and various fevers; fleas with their typhus; and biting flies with their sleeping sickness and river blindness. There are also reduviid bugs, including the bed bug's distant cousin *Triatoma infestans*—also called the kissing bug, one of the species used in the US Army's failed Vietcong detector— which spreads Chagas disease from the southern United States down through South America by passing the tropical parasite *Trypanosoma cruzi* to hosts as they sleep. Left untreated, which it often is because the disease is difficult to detect in its early stages, Chagas causes chronic and typically fatal heart problems. Other close bed bug relatives can also spread disease, although not to humans. For example, *Oeciacus vicarious*, the swallow bug—which mostly lives in the feather-lined mud nests that cliff swallows dab onto steep, rocky bluffs—is able to spread Buggy Creek virus and a handful of other illnesses between birds.

Why should these insects spread disease and not the bed bug?

Perhaps it was never in the right place at the right time to form a tight bond with a pathogen. Most insect-microbe relationships establish over ages, long before humans even entered the picture. The kissing bug and *T. cruzi*, for example, have a history that may go back 40 million years; their relationship was humming along efficiently when humans became inadvertently entangled in it at least 9,000 years ago. By coincidence, the kissing bug has unique characteristics that make it particularly good at spreading the parasite to people, including the ability to take in a sizable meal. The *T. cruzi* virus is found at very low levels in infected blood, and most insects aren't big enough to take the amount of blood necessary to pull up the parasites. The kissing bug also eats so much in a single bite that it's forced to defecate mid-meal. Its feces, which are laced with the virus, end up on a person's skin, where it may get scratched into the bloodstream. The bed bug, on the other hand, doesn't appear to have developed a long-term relationship with *T. cruzi* or any other pathogenic microbe.

Another factor that may limit the bed bug as a vector is that our relationship with it, unlike our interactions with most other biting insects, is intimate. Vector insects typically have vast territories; mosquitoes or biting flies may fly several miles to find a meal, and fleas can spring more than thirteen inches in a single leap from one host to another. These insects often feed on multiple humans as well as other animals; each of us is a single taste from a smorgasbord of blood. The more unique meals the insects take, particularly in disease-ridden regions, the higher the chance of spreading a pathogenic microbe. While the common bed bug is capable of travel, it can't go far on its own volition. For one bed bug to feed on multiple hosts, it must hitchhike to a new location or live in a place either jammed with many people or with a consistent turnover. Instead, most individual bed bugs live in a single home or room and feed off a small number of human hosts—possibly just one— throughout their life cycle. The chance of a single person infected with a transmissible disease being bitten by a bed bug, combined with the chance of that bed bug ingesting the microbial cause and then infecting a new host, becomes very small in this scenario.

Yet another hypothesis suggests that the bed bug actually *is* capable of spreading disease. But rather than a true biological transmission, where a microbe grows or multiplies within the insect before it passes on to an unwitting human host, this hypothesis suggests that bed bugs are capable of transmitting disease mechanically (the MRSA and VCE transmissions, had they been true, could theoretically happen this way). A bug might very occasionally have viable virus or bacteria cling to its mouth or other body parts. From there, the bug may rub the pathogens into a person's bodily fluids or blood, perhaps through an open wound or the absorbent mucous membranes lining a body's openings to the outside world, including the mouth, eyes, and nose. From there, the microbes could enter the bloodstream and cause an infection. So far, the idea remains unproven. But if the bed bug's limited host selection truly impacts its ability to spread disease, it may be moot. Microbes typically don't live outside of a host's body for very long. The chance of survival on a bed bug's mouth are slim, particularly considering the amount of time it may take for the bug to be whisked away to a new host. An exception could be hive-like dwellings such as apartment buildings, where the bugs may come into contact with more people over a shorter period of time. Still, there is no evidence of bed bugs causing outbreaks in these types of living spaces.

It may be that no one has found that bed bugs cause disease because they haven't been looking in the right places. That is the hypothesis of entomologists at Virginia Tech, who in 2013 argued that most viral pathogens tested on bed bugs aren't known to spread by any other bloodsucking insect, either. There is no record, for example, of HIV ever successfully and regularly existing inside a biting insect. Since this and many other viruses don't have a long history with insects, or experience living and replicating inside of them, it's unlikely that the viruses would take a sudden evolutionary leap to bed bugs.

Instead, the Virginia researchers argue, it'd be more likely for a bed bug to spread a virus that has already been living in another animal, like a bird or a bat, for a very long time. Such a virus may have evolved to survive in the bugs and to be transmitted from bird to bird or from bat to bat. Then, if humans interrupted the cycle, just as with the kissing bug and the *Trypanosoma cruzi* parasite, we may contract an as-yet-unknown illness.

. . .

Even if bed bugs aren't able to spread disease, they aren't harmless. For many people, there are the immediate effects of a bite on the skin. As one early twentieth-century dermatologist put it, the bugs "can produce a good deal of cutaneous mischief." A bed bug's tiny mouth is too small to leave an obvious mark; it slides effortlessly into the skin like a tiny hypodermic needle. But the proteins in its saliva can spark allergic reactions. Depending on the person who is bitten and their unique immune system, these reactions may manifest as itchy welts or rashes.

More severe cases erupt in blood blisters. Once, at an entomology conference, a researcher from the US Department of Agriculture sidled up to me in a crowded auditorium. In between presentations, he told me about his own odd bed bug case. He had been unintentionally bitten during a bed bug experiment in 2009, and he came down with cutaneous vasculitis, a disorder that causes inflammation and hemorrhaging in the skin's small blood vessels. Prior to that accidental bite, the scientist told me he had been knowingly exposed to bed bugs more than thirty years earlier when he offered to help a fellow graduate student at the University of California, Davis, feed bugs and lice in the lab. Back then the bugs just gave the scientist a rash. In the more recent case, he theorizes that his immune system still remembered the allergens in the bed bug's bite so many years later, prompting the fast and painful response. He is likely right. In some cases, allergies from insect bites—as well as foods and drugs—get worse over time as the immune system launches increasingly heightened and targeted attacks against them.

Bed bugs may also trigger asthma or, extremely rarely, anaphylaxis, a potentially deadly allergic reaction that seizes the entire body. The itch from a bite may drive a person to scratch, opening their skin to a variety of microbes and causing an infection—the same problem that sent me to the emergency room from bed bug bites nearly a decade ago. And despite the small amount of blood the bugs are able to consume at each meal, severe infestations where a person is bitten hundreds or thousands of times daily can accumulate in enough blood loss to cause anemia.

Even people who voluntarily feed the bugs may face health

problems. When I watched Harold Harlan strap a box of bugs to his arm in 2011, he described to me a period where he had a peculiar systemic reaction, which he suggested might be linked to feeding his bed bugs twice as often as usual in order to increase the population. "I had a sort of malaise and was a bit feverish," he said. "And I had a bit of an unusual stomach upset." When he cut back on the number of feedings, the symptoms vanished. Similarly, in his monograph, Usinger wrote that during his seven years of feeding bed bugs, his hemoglobin dropped to below-normal levels. (Hemoglobin is a protein in red blood cells that takes oxygen from the lungs and delivers it throughout the body.) This condition persisted even when Usinger took both iron pills and injections, and only improved when he stopped feeding the bugs altogether.

With such varied reactions to bed bugs, it isn't surprising that doctors have difficulty in diagnosing the insects' bites. Still, this is a problem for patients who come in for help: for example, when Anoki's dermatologist couldn't confirm whether his rash came from bed bugs, or when my own doctor unnecessarily tested me for Lyme disease. Some cases are far sadder and show worrying signs of medical neglect, most typically among the elderly, disabled, or the very poor—which are the demographics most vulnerable to bed bugs.

One story comes from a woman I'll call Shirley, whom I met in 2012 while tagging along with Dini Miller and her team from Virginia Tech as they tested bed bug prevention strategies in a subsidized housing development in northern Virginia, just on the edge of Appalachia. While most of the apartments we visited were clear, Shirley, a seventy-two-year-old retired hospital housekeeper, had bed bugs. It was the apartment complex's twenty-second bed bug case over the previous year and a half.

As soon as we stepped into her apartment, Shirley, who was dressed in pink Daisy Dukes and had lipstick escaping from the edges of her mouth, showed us the swollen bites on her forearms and legs. She had scratched the welts until they bled. "Look it there, how they bit me," she said, her accent like pulled taffy, adding that the bugs had been bothering her "a good long while." As she and I talked, Miller's team inspected her cramped bedroom. Among the knickknacks scattered around—including a plastic New Year's

Eve tiara, artificial flowers, and a signed and framed photograph of Shirley's favorite local wrestler—Miller's technician and husband, Tim McCoy, found an orange and white tube of cream. It was 5 percent permethrin, a topical pyrethroid used to treat scabies, which are tiny mites that burrow under the skin. Her doctor had prescribed it, assuming that her swollen arms and legs were from this itchy arthropod rather than bed bugs. And so, she told us, she had been rubbing the insecticide—a concentration ten times greater than the amount a pest controller would be allowed to spray in her bedroom, according to Miller—into her bleeding sores. Miller searched for the doctor's phone number and had a heated exchange with his assistant, insisting that Shirley had been misdiagnosed and needed a different treatment. By the end of the day, a package of hydrocortisone cream showed up at Shirley's apartment. But she also kept the permethrin. When the Virginia Tech team checked on her the following week, they learned she had used the entire tube.

• • •

The real and perceived physical threats from a bed bug infestation pale next to the impact on mental health, which reaches into far darker territory than Anoki saving bed bugs in his freezer. In 2012 a medical and veterinary entomologist from Mississippi State University named Jerome Goddard, a tall and booming man who has jokingly sparred over bed bugs with Comedy Central's Stephen Colbert, and Richard deShazo of the University of Mississippi Medical Center published a survey of 135 blogs and online forums on bed bug infestations. The two researchers were interested in uncovering patterns of psychological distress from the bed bugs, which they hypothesized were similar to those seen in posttraumatic stress disorder. Their survey revealed that more than 80 percent of the bed bug posts showed psychological damage including "nightmares, flashbacks of the infestation, hypervigilance (to keep the bugs away), insomnia, anxiety, avoidance behaviors, and personal dysfunction." Only one of the bed bug posts in Goddard and deShazo's study scored high enough on a standard PTSD test for its author to be considered as having the disorder, but several matched at least portions of the list. Most alarmingly, five of the posts revealed thoughts or attempts of suicide.

One of the reported symptoms, insomnia, may exacerbate the others, forcing the mental reaction to bed bugs into an unfortunate positive feedback. ("We all know the effects of sleep deprivation. And what causes more sleep deprivation than a bed bug?" an entomologist once said to me.) Indeed, sleeplessness may spring from the anxiety of getting bitten at night, the money and time spent on treatments, or the itch of the bites. This can lead to days, weeks, or months of interrupted sleep, which may continue even after the bugs are gone. No matter the cause, sleep deprivation can lead to everything from feeling anxious in social situations to having trouble focusing or reacting; the latter problems feed into at least 100,000 car accidents in America each year as well as major disasters including the Chernobyl nuclear accident and the Exxon Valdez oil spill. Seen in this context, a world full of zombified bed bug victims is unnerving.

A random selection of 135 blog posts pulled from the Internet is not necessarily representative of all bed bug sufferers, especially since these posts are self-reports. There is no way to confirm whether they are true. Still, other experiments and case studies support the idea that the bugs can cause deep psychological trauma. In 2012 public health and housing specialists in Montreal published a preliminary study on ninety-one tenants living in two apartment complexes that were under investigation for unfit living conditions. Thirty-nine had confirmed bed bug infestations, and the remaining units did not. The researchers gave the tenants standard tests that measure anxiety, depression, and quality of sleep. Compared to those who had no bed bugs, the people with infestations were far more likely to suffer insomnia and general anxiety. They also tended toward symptoms of depression, although here the study was inconclusive. These symptoms occurred even for people who had no previous history of either anxiety or depressive disorders. (When I read this study, I remembered something my internist told me during a checkup as we chatted about my book: "Sometimes, I have to sedate people who think they have bed bugs. I give them Valium.")

Also in 2012, a group of psychiatrists in New York chronicled the bed bugs and resulting trauma experienced by seven psychiatric patients from the city's vast, unwieldy hospital system. In two cases, the doctors concluded an "impaired quality of life and psychosocial functioning," where bed bug infestations caused already-vulnerable people to lose their social safety net. One, a paranoid schizophrenic in her late forties living alone in a housing project, was kicked out of her church after she apparently left bed bugs in the pews, and her boyfriend stopped coming to see her at home. The apartment of the other, a man in his early fifties with schizoaffective and bipolar disorders, was in such disarray from bed bugs and other problems that his social worker refused to visit him.

Social shunning isn't limited to people with psychiatric diagnoses or those who are in the social services network. Friends and interviewees have admitted to me that because of bed bugs, they've broken up with significant others, lost friends, avoided social interaction with friends and family, and had those interactions denied to them. Still others, on their quest for dates or one-night stands during the glory of their singlehood, view bed bugs at a hook-up's home as more worrisome than a sexually transmitted infection. Collectively, these people have told me that the bed bugs made them lonely and ashamed, that researching remedies made them feel manic, and that cleaning and preparing for a professional pest treatment made it seem like they were losing their grip on reality. When I visited a housing project in central Ohio, an elderly disabled man living there told me about the bed bugs that had plagued him for four years, affecting his ability to see his grown sons and grandchildren. I asked how the bugs made him feel, he looked at me with watery gray eyes and replied: "I got one of the sorriest feelings."

The New York case studies also described two patients with bed bug delusions, despite having no evidence of an actual infestation or even any insect bites. One, a man in his thirties, couldn't shake the idea that he needed to clean his home with bleach every day, even after six weeks of treatment and antipsychotic medications. The other, a seventy-five-year-old woman, would bring pieces of lint and dust stuck to cards or clear tape to the hospital; she thought these fuzzy specks were bugs. This behavior extends beyond bed bugs to any imaginary insect and is common to a psychotic disorder called delusory parasitosis—also called Ekbom syndrome for the Swedish neurologist Karl-Axel Ekbom, who first published accounts of it in the late 1930s—in which patients think insects, worms, or mites are burrowing into their skin. Often a person with delusory parasitosis will think their particular parasite is new to science, which helps them reason away the doctors, entomologists, and pest control inspectors who insist that they are not infected or infested with anything. When one expert tells them the bugs are not real, they send jars or envelopes of dust and lint to another.

The delusional patients may go so far as to dig into their skin with scissors or tweezers to pull out their imaginary torturers. Salvador Dalí, a longtime entomophobic, suffered such a self-mutilation during what would likely be categorized today as a bout of Ekbom syndrome. In his autobiography, *The Secret Life of Salvador Dalí*, the surrealist painter recalls seeing two or three insects on the ceiling of a Paris hotel room as he lay in bed recovering from tonsillitis. He fell asleep, and when he awoke, there was only one insect left. Convinced the others had fallen on him, he frantically searched his body until he found what he thought was a tick or a bed bug attached to the back of his neck. He jumped up to inspect it in the mirror. He could see the bug in the reflection but could not pull it off—he squeezed through with his fingernails until he drew blood, but it still wouldn't move. He wrote: "It was as if it was formed of my own flesh. As if it constituted an inherent and already inseparable part of my own body; as if, suddenly, instead of an insect it had become a terrifying germ of a tiny embryo of a Siamese twin-brother that was in the process of growing out of my back, like the most apocalyptic and infernal disease." Then he grabbed a razor and chopped it off. The hotel chambermaid found him covered in blood and called the manager, who called a doctor, who told Dalí that the "bed bug" was in fact a small birthmark.

The final three case studies in the New York psychiatrists' paper told the sad tales of three young women in their early twenties, all with actual bed bug infestations that exacerbated or sparked underlying psychiatric problems. (Two had bipolar disorder, one with a history of substance and childhood abuse, and the third had no history of mental illness but was ultimately diagnosed with acute adjustment disorder as well as anxiety and depression.) For all three women, the bed bug infestations caused a severe mental breakdown, which isn't entirely surprising. Bed bugs may be difficult to deal with even in the best of circumstances; for those in fragile mental states, the bugs are devastating. For two of these women, it was nearly deadly. The young woman diagnosed with acute adjustment disorder tried to kill herself by getting drunk and then taking twenty-two acetaminophen tablets because she was afraid she would never be able to function normally again. And one of the women with bipolar disorder, as her apartment was being treated for the bed bugs, formulated what would be an unfulfilled plan to jump off a New York City bridge.

Not everyone escapes the pull of suicide. A 2013 report from Montreal describes a sixty-two-year-old woman who had bipolar disorder and borderline personality disorder. She was also an alcoholic, a chronic gambler, and had attempted suicide three times since she was in her mid-twenties. After having two bed bug infestations over a period of about six weeks, during which she relapsed on both her alcohol and gambling addictions, the latter of which lost her a lot of money, she woke in the middle of the night to find a drop of blood on her nightgown. She wrote a note that the bed bugs were back and that she wanted to go to a better world. She also sent an e-mail to a friend: "I am panicking now because I just saw a drop of blood on my dressing gown sleeve and I am sure that vampires are back and I cannot stand to live in fear of me being eaten alive. . . . I cannot stand it and I chose to take my life. . . . At the time of writing, I have swallowed a bottle of wine and two hundred pills and I feel nothing, I feel completely empty, it is unbearable. . . ."

She called her boss to tell him she would not be at work the next day. Something in her voice made him call the police. She stepped out on her balcony on the seventeenth floor of her apartment building. The police arrived but could not coax her down. She jumped.

• • •

As I learned about these increasingly wrenching bed bug tragedies, I considered my own mental state during my bed bug quest. Was it normal to dig so deeply? To try to find out where these bugs came from and why they existed? What was wrong with me? Rather than exterminating my own bed bugs and trying to move on, I had shifted in the opposite direction toward a deep fixation.

But it wasn't quite the same obsessive quality as being convinced that the bugs were still there, or that I might pick them up in any number of public spaces at any given time, or that nothing would ever be good again. Instead, rather than looking away or living in a state of constant worry, I looked for more—more stories, more science, and even more songs and art. I had started out my project simply wanting to learn about the bed bug and its story. But at some point, the act of collection had become compulsion. It was as though knowing everything I could possibly know would help me conquer the beast. Where was all of this headed?

One evening I was working late, sitting at my desk with a blank screen and a blinking cursor in front of me. Harlan's gift, the dead bed bugs in a vial of alcohol, sat under my computer monitor, as did a growing collection of plush bed bug toys and a custom rubber bed bug stamp that a friend had given me at my bridal shower. Usinger's autobiography and several tattered notebooks were open on the desk. I thumbed through them, skimming the stories and notes, willing a pattern to leap from the pages.

I wondered again about the bed bug's two origins—both the very beginning when it first emerged in bat caves hundreds of thousands of years ago and the moment a resistant strain seemed to pop up all over the world over the past two decades. It seemed impossible to untangle the way each appearance had happened. But maybe I could pick one or two possible origin stories and try to follow them to their conclusions. I read through more notes, particularly on the theories of the bed bug resurgence and its hypothetical sources somewhere in Africa or in Eastern Europe. I thought of Usinger's travels in Kenya, South Africa, and the Congo, as well as his unfulfilled trip to the former Czechoslovakia. I put down the notebooks and started searching through my contact lists and pricing airplane tickets.

SEVEN

MONEY

The Wild West of the Bed Bug Economy

On the stage of a cavernous ballroom on the fifth floor of the Las Vegas Red Rock Resort, a balding man in a red golf shirt bounced with the excitement of a bulldog puppy as he welcomed the audience to the 2012 BedBug University North American Summit. More than four hundred people had traveled to the conference, which was in its third year, to talk about bed bugs and to network. I had joined them for insight on just how big the bed bug economy had become. A couple hundred of us sat expectantly at long conference tables draped with white linens, our eyes on the man on the stage. There were pest control operators, research entomologists, and representatives of the industries that were carrying the brunt of the bed bug resurgence, including hotels, summer camps, and colleges. The energetic man in red outlined the two-day schedule, which was packed with lectures and socials. Then he introduced one of the summit's first speakers. Music crackled though the loudspeakers. It was LMFAO's "Party Rock Anthem."

> *Party rock is in the house tonight*
> *Everybody just have a good time*
> *And we gonna make you lose your mind*
> *Everybody just have a good time*

The speaker—a tall man with an impeccable goatee, glasses, and a tailored suit—climbed the stairs of the stage, his face a mixed grimace of amusement and embarrassment. He pulled up a set of presentation slides, which flickered onto two large screens bookending the stage, and fiddled with a collar-clip microphone.

"I guess this is why . . . ," he said,

We just wanna see ya . . .

". . . you don't agree to a musical introduction . . ."

. . . shake that . . .

". . . after a few drinks the night before your talk . . ."

In the club party rock, look a pretty girl
She on my jock (huh) non-stop when we in the spot
Booty on the way like she owns the block.

The music abruptly shut off. As the chuckles from the audience faded, the presenter, a British entomologist from the University of Sheffield named Michael Siva-Jothy, gestured to his first slide. It read: "Know Your Enemy: Why Pure Research Is Relevant for Bed Bug Control." Siva-Jothy explained that his team of evolutionary and ecological entomologists at Sheffield had been working with bed bugs since before the resurgence in the UK, originally drawn not by the mystery of their comeback or possible new methods to get rid of them, but by their unusual mating habits and immune systems.

In the female bed bug, the reproductive system doubles as an immune system: the male's violent stabs cause infection, which the female wards off with her major reproductive organ, the immune-cell-filled sac called the spermalege. And for both sexes, a mixture of bacteria both on and in their bodies requires robust, unique defenses. But since the resurgence, understanding bed bug behavior, immune systems, and mating patterns could eventually help combat the pest, which is why so many exterminators had packed into the ballroom.

After a brief introduction to bed bug sex, Siva-Jothy took the room through a series of experiments his team had done to unravel the bed bug's odd sexual and immunological mechanics. In one test, the scientists placed a male bed bug in a small flat dish lined with filter paper, gallantly named a "mating arena." Here, the scientists paired the male bug with either a recently fed female or an unfed one to see which he preferred. Later the scientists ran the test again but replaced the sated female with one they had stabbed with a sterile syringe and pumped full of air in order to see whether her full figure was the seductive factor or if something else was at play. The scientists ran dozens of trials, switching which female they fed, starved, or pumped with air to control for any bias the male might have toward a particularly attractive mate. They found that the males preferred the fatter females, even when that fullness was artificial. The males were also more successful in mating with the plump females; the unfed ones could press their flat bodies against the bottom or sides of the mating arena to shield their abdomens while their round peers could not.

Siva-Jothy advanced through the slides to another series of experiments in which his team killed a male bed bug, broke off his penis, and smeared it on a plate of agar to see what types of microbes lived on it. Analysis showed the penis had been home to fungi, specifically two species of *Penicillium*, relatives of the organism that produces the drug penicillin, as well as bacteria in the genera *Arthrobacter*, *Bacillus*, *Enterobacter*, *Micrococcus*, *Staphylococcus*, and *Stenotrophomonas*. A similar test found an additional four types on bed bug exoskeletons, and the same microbe species were also found in samples taken from the containers where the bugs lived in the lab.

The next slides described how the scientists stabbed groups of female bed bugs with either a sterile needle or one dipped in a soup of bed bug penis microbes. Siva-Jothy's team stabbed some of the females in the spermalege and the rest on the opposite side of the abdomen, which has no protective sac. The sterile needles didn't cause damage no matter where they stabbed. The dirty needles, however, caused the worst wounds on the side without the spermalege. The organ, it seemed, is like a built-in prophylactic against bed bug STDs.

The final slides described research on general bed bug behavior. For these experiments, the scientists placed bed bugs in a large enclosure, where the insects had freedom to eat and take refuge at their own will. Bed bugs normally live as close to their food source as they can. But in a series of tests, the researchers found that the less complex the structure was around the feeder, the more likely the bed bugs would seek shelter somewhere else, even it if was farther away. In other words, smooth surfaces that offered fewer places to hide seemed to drive the bugs to more comfortable locations.

As I typed notes on my laptop, I wondered how any of the experiments might lead to new pest control methods. In the future, would exterminators release dozens of artificially plumped female bed bugs in an apartment to seduce males away from the females who were actually fed and therefore capable of laying eggs? Or spray beds with liquefied symbiotic bacteria, genetically engineered to hurt the bed bugs' immune systems or attack their reproductive organs? Or could changing the complexity of a bed's construction make it easier to find the bugs?

Siva-Jothy flipped to his final slide, and BedBug Summit employees dressed in red golf shirts ran microphones around the room for questions from the audience. One entomologist asked whether carefully balancing the male-to-female ratio in a laboratory colony would prevent the males from stabbing the females to death. Siva-Jothy suggested it might not matter. A pest controller wanted to know if it was true that female bed bugs run to the corners of a room after a meal to avoid sex. "No," Siva-Jothy replied, although several other exterminators in the room shook their heads at this.

One of the last questions came from Mike Potter from the University of Kentucky, who asked whether the bed bug's apparent aversion to simple structures might encourage bed bugs to disperse from form-smoothing mattress and box spring encasements. I thought about the zippered covers on my own bed at home.

"I think you're right," Siva-Jothy answered, "assuming casing is having the desired effect is too simplistic. There are all kinds of complicated stuff going on here. It may well have some beneficial effects, but I suspect it has unforeseen consequences."

• • •

After the talk, I wandered across the hall to an equally large ballroom, where around forty-five vendors had set up in long rows of booths. Each paid, at minimum, $1,000 to be there. As I walked past each station, I watched salespeople lure potential customers with candy, plush bed bug toys, and key chains. I saw a bed bug DNA swab kit; several canine units including a bed-bug-sniffing Labrador mix available for live demonstrations; the modern plastic molded versions of the saucers our great-grandparents placed under bedposts a hundred years ago; the 25b sprays boasting

A pile of plush bed bugs (actually dust mites passed off as bed bugs). Credit: Brooke Borel.

essential oils and other "All Natural!" ingredients; insecticides from chemical giants including Bayer CropScience, FMC, and BASF; computer software to track business, find customers, and advertise; a bed bug pest control franchise offering ten-year contracts; a precision duster to shoot diatomaceous earth or powdered poison into cracks and holes; electric and propane heating systems; the Entomological Society of America; a vacuum cleaner with a specialized filter; two pest control trade magazines; a self-heating suitcase prototype I would eventually see advertised in *SkyMall*; and, of course, mattress and box spring covers.

I paused at the booth of a business that advertised not a bed bug killer, detector, or educational service, but the vanity phone number 1-800-BEDBUGS ("The Number No One Forgets"). Months later, back home in New York, I recognized it on the subway, plastered on posters running the lengths of the cars and strategically placed at eye level. I asked the friendly bearded man at the booth, who introduced himself as Michael Eisemann, about the number.

In the nineties in Detroit, Eisemann said he had been working as a commercial real estate manager. In 2009 he was ready to take the step from management to ownership, so he spent nearly all of his savings on his first building, a 47-unit apartment complex in the city's New Center neighborhood. It seemed like a good investment. Thanks to the financial crisis, the owners were forced to sell at a quarter of what they paid. Eisemann planned to turn the building into affordable housing. But two weeks after closing, he got a phone call from his building manager, who, in Eisemann's memory, asked: "So what are you going to do about these bed bugs?"

Around the same time, the city of Detroit had organized a bed bug task force made up of local business owners, tenants, entomologists, and members of the city council. Eisemann attended and recalls an angry crowd. The people at the meeting said the landlords were negligent, practically infesting the apartments with bed bugs. Someone shouted "lawsuit," and Eisemann went numb, imagining his new investment crushed by an avalanche of legal disputes.

But as he sat there listening to the heated exchange, he had an idea. Eisemann had a business selling popular 1-800 numbers since around 1993. *You know what would be a great vanity number?* he thought to himself. *1-800-BEDBUGS*. If he had the rights, he could license it nationwide, and one lucky pest control operator in each region would get to use it.

As soon as Eisemann left the meeting, he began to hunt down the number. But a Fortune 500 company already owned it, an international service company that Eisemann declined to name as we chatted at his booth. The company wasn't using the digits to intentionally spell "bed bugs." Instead, 1-800-233-2847 was one of many randomly generated phone numbers that the company had bought in bulk. Eisemann went straight to the international head of marketing to negotiate the purchase, which took six months and a price he wouldn't disclose to me.

Since buying 1-800-BEDBUGS, Eisemann had licensed it to pest control operators in between fifteen and twenty regions across the United States, including New York, Chicago, Cincinnati, and Alabama, and claimed that it boosts their advertisement response between 20 and 40 percent. A major chemical company, which he also wouldn't name, offered him a million dollars for the number, which he turned down because he "believes the number has greater potential." I pressed for hard figures, but he demurred, brushing off my questions on how much the number makes or whether it is more lucrative than the four buildings he now owns in Detroit. He'd only say it "does very well" and that owning it is not so different from his role as a landlord. "Licensing a number is like renting apartments, only without the headache of the toilets. Or the bed bugs."

I left Eisemann and wandered through the rest of the booths, snapping photos and taking notes. A group of pest control operators had congregated in one of the aisles, and I stopped to talk with them. When they found out I was writing a book about bed bugs, they whipped out iPhones to show me images and videos of the worst infestations they'd ever seen. One gnarly set came from the Manhattan apartment of a partially blind man. "He thought it was mold," explained the proud owner of the photos as he flicked across the screen of his phone and spread his thumb and forefinger to zoom in on a fitted bed sheet darkly discolored around the edges. It did look like black mold, but in fact it was bed bug feces. The other exterminators who had joined the group groaned and then tried to one-up him with their own photos.

During the summit, I would also meet a woman who works for the Las Vegas hotel industry who said she wished DDT would make a comeback; three people who casually asked if I'd read their own bed bug books as they pressed business cards into my palm; a pest controller from Long Island who told me that a nervous hotel manager once asked him if he'd been trailed by the press; and an employee of the Texas-based company the Bug Reaper, who showed me photos of the conspicuous, refurbished yellow hearses his team drives. Inside, canisters of insecticide are concealed in caskets.

• • •

None of this networking would have been possible without Phil Cooper, the creator of both the summit and the entity that runs it, a small northeastern company called BedBug Central. Cooper, the energetic red-shirted man who opened the conference with "Party Rock Anthem," is the CEO of Cooper Pest Solutions, where he works alongside his brother, Rick Cooper, the same reserved en-

The exhibitor ballroom at the 2012 BedBug University North American Summit in Las Vegas. Credit: Brooke Borel.

tomologist who pleaded with academics to study bed bugs in the early aughts.

In 2007 the Coopers had a sense that bed bugs weren't going to go away. To prepare for the inevitable onslaught of business and requests for information, Phil Cooper started a website. Every Sunday morning for three years, he sat at his kitchen table with a web developer and ultimately built BedBugCentral.com, a clearinghouse for product reviews, general information, and DIY pest control projects.

By 2009 BedBug Central was an official subsidiary of Cooper Pest and hired its first full-time employees. The same year, as the media frenzy flamed the fears of the general public, the site shifted to selling bed bug products, highlighting items and vendors that were part of what Phil Cooper now calls the "BedBug Central Kool-Aid," or the system he thinks works best. He also signed up for Google AdSense, which allowed any company to advertise on his site, including those with products that BedBug Central didn't endorse or even like.

Also in 2009, the company started BedBug University, its educational branch. This included a $2,000 per person hands-on bed bug management boot camp at Cooper headquarters for pest control operators, shorter national road shows that cost a few hundred dollars, and the BedBug Summits like the one I attended in Las Vegas. By this time, the company was also under way with the BedBug TV YouTube channel, which within four years would post eighty-seven videos that had collectively amassed nearly 1.5 million hits, including one on how to check a bed for the bugs, which users had clicked on more than half a million times.

By September 2010, BedBug Central launched the first and largest American conference dedicated solely to the bed bug. The Coopers had expected 225 people. More than three times as many signed up, and most had to go on a waiting list. The inaugural BedBug University Summit ultimately crammed 360 people into a hotel conference room just outside of Chicago, and they were "practically hanging from the chandeliers," Phil Cooper has told me more than once. BedBug Central issued around seventy press passes. *The Early Show* and *Good Morning America* had live national feeds, and the event made the front page of the *New York Times*.

When I visited the summit in Las Vegas two years later, the market had matured and media attention had waned. Or at least that's how Phil Cooper saw it. Competitors would tell me that the drop in numbers corresponded to a fall in BedBug Central's popularity. Either way, after swelling to 650 people in its second year, attendance had dropped by 30 percent. Only six members of the press were there, including a few local news affiliates, the *Las Vegas Sun*, and me. Still, the four hundred attendees paid $595 for a ticket and nearly double that if they had a booth. Silver, gold, and platinum sponsors shelled out $5,000, $7,000, and $10,000, respectively, to have their names advertised on swag bags filled with programs, and pens and notepads emblazoned with "BedBug Central," or on the black and red name tags everyone wore, or on the programs themselves for sponsoring one of the continental breakfasts, the beer tasting, or the "Night with the Experts," a mixer in the hotel's dramatic lobby bar where scientists and other speakers handed out drink tickets.

It was at the mixer that I learned of Siva-Jothy's plans to travel to

Kenya to study a bed bug relative living in mountain caves on the border with Uganda. The bugs were a type of bat bug with a reproductive system unusual even by cimicid standards—males sometimes masqueraded as females with fully developed spermaleges. This made for an interesting experimental subject for Siva-Jothy's team. He and a few colleagues had traveled to Kenya several times before, initially using descriptions in the *Monograph of Cimicidae* to track down bat caves scattered all over the country, which Usinger had visited during his own research decades earlier. I started fidgeting in my chair. Perhaps this was my way into Africa, to trace some of Usinger's path. I asked a few questions and then listened patiently, waiting for the right moment. During a pause, I broke in with what I hoped was a casual tone and asked Siva-Jothy if he'd consider letting a writer join his upcoming trip. He gave me a long stare and asked if I'd ever been caving. I had not. He asked if I was claustrophobic. I wasn't sure. He waved me off with stories of caves full of elephant carcasses and hundreds of thousands of bats, and then changed the subject.

The group slowly dispersed, and I wandered around the bar. As the alcohol flowed, I overheard bits of conversations—researchers and pest control people candidly swapping bed bug stories and dishing on the business. But the scientists and industry people don't always mix so merrily. Unlike an academic meeting or a commercial trade show, this conference coarsely blended science and industry. The scientists had no restrictions on what they presented and reported data showing that products for sale in the ballroom across the hall didn't work. BedBug Central also didn't require the scientists to disclose potential conflicts with the products they reported on. Some researchers invented the very technology they raved about, and many had received direct funding from companies that sell chemicals or other products, although there was no obvious mention of the relationships in their presentations. (Nearly every American bed bug researcher I spoke with regularly takes money from the big chemical companies, which goes toward testing the companies' pesticides and supporting other research, in part because of scant available federal funding for bed bug work.) Some vendors had no scientific basis for their products, and critics questioned BedBug Central's intentions for allowing these companies in, hinting that even bad merchandise brought in money in the form of conference fees.

Other products posed knottier problems centering on the bed bug's complex behavior. Most researchers and exterminators agree that bedding encasements protect at least part of the bed from the moldy appearance of bed bug stains. The covers also make treatment easier by sealing off potential hiding places, such as the cracks between exposed wooden planks and under screws on the underside of a box spring. But Siva-Jothy's presentation showed that these artificially smoothed surfaces might make the bed bugs seek a better crack somewhere else in the room. Adding to the complications, Cooper Pest has a long-standing relationship with Protect-A-Bed, the major encasement brand on prominent display at the summit, and even helped design their product. (Phil Cooper claims he isn't a shareholder but wouldn't discuss the relationship in detail.) The Las Vegas ballrooms were full of loose threads, I realized, a truth even in such a niche of economy. Pull one thread, and nearly everyone's conflicts of interest were exposed.

BedBug Central's business strategies and drive for publicity have not gone unnoticed by competitors. Weeks after the Vegas summit, I would receive an unsolicited e-mail from one of them, a man who was angry with me for sharing on Twitter a bed bug segment from the Animal Planet television show *Infested*. The star of BedBug TV, Jeff White, appeared in the clip. "Jeff White and Bed bug central certainly think they are at the center of the bug bug industry [sic]," the sour pest control operator wrote. "They are in fact just a company bundling anything they can sell, promote or make a commission on with in the bed bug industry [sic]. The Walmart of the Bed bug community [sic]." He added that his own heat treatment service has a 99.125 percent success rate. It was too bad I'd never see it, because he also rescinded a previous offer to let me visit his company, and then accused me of being paid off to write the Twitter message (I was not). I stared dumbfounded at my computer screen.

The e-mail message was nothing compared to the Coopers' most vocal critic. If BedBug Central is the Walmart of the bed bug world, then David Cain, a bed bug pest controller in London, is its Michael Moore. Cain is an imposing figure with icy-blue eyes who is invariably dressed in a black golf shirt, black leather pants, and tall black work boots. In a smoky casino lounge attached to the conference center where the BedBug Summit took place, he told me his manner of dress was inspired by Albert Einstein, who, Cain claimed, wore the same clothes every day because he didn't want to waste his brain power on picking out different outfits. That, and Cain could stomp his heavy boots after leaving a bad bed bug infestation and the bugs would simply slide off his legs.

Cain's pest control business, Bed Bugs Limited, specializes solely in bed bugs. He told me he has treated 24,000 cases in more than ten years, and that he has a savant-like ability to detect a single bed bug in a room. Cain also runs his own bed bug education website, BedBugBeware.com, which has been online since 2007. On the surface it seems like the British equivalent of BedBug Central and perhaps even a direct competitor. Make that suggestion to Cain, I'd soon learn, and risk a brusque lecture.

Cain has fashioned himself as a bed bug vigilante, defining all he encounters in the business into rigid categories of right and wrong. And to him the Coopers' moneymaking approach falls squarely into wrong. In London his bed bug treatments run an average of £140, or around $190, a fraction of the potential thousands Americans might pay (one wealthy family on Manhattan's Upper East Side reportedly spent $70,000 clearing their bed bug infestation, the highest figure I've found for a private home). On the Internet, in addition to the product testing that Cain publishes on his own website for free, he is the second most prolific poster on the forum BedBugger.com, where he has posted more than ten thousand times on products and services and given free advice to pretty much anyone who asks. Each year at the BedBug University Summit, he polices the academic sessions by disclosing scientists' patents and other perceived illicit ties during the Q&A's or arguing with people in the hallways. In Las Vegas, Cain told me that Phil Cooper assigned BedBug Central employees to keep tabs on his conversations, an accusation that Cooper denied.

Cain didn't let me off the hook, either. The first time we met, during the "Night with the Experts" party, he delighted in telling me about the bed bug museum he has been curating at his London business—a collection of art and books and odd memorabilia.

In our second conversation, after learning that I was the author of an article he did not like, he told me he was writing an official report condemning it. At the summit and over several months following it, Cain would also give me a complimentary sample of his patented bed bug monitor and a copy of his bed bug book, tell me portions of his life story, mock me in e-mails, blind copy me on e-mails to other people in which he mocked them, submit a rude entry describing an enemy's name to Urban Dictionary (an online repository of definitions of slang and off-color phrases), insist on both hello and good-bye hugs when I interviewed him over lunch at an Ecuadorian restaurant off the Long Island Expressway in Queens, and end conversations abruptly. He would also reveal that he squats on more than 350 bed-bug-related domain names, which he refuses to give to bed bug businesses unless they pass his ethics test.

According to Cain, BedBug Central is a "bed bug cartel." He pays to attend each of its summits and others like it worldwide, he said, to keep an eye on what people are doing there. To keep them honest, or at least to let *them* know *he* knows they are not. ("If I disappeared, the corrupt idiots would think they that they'd won," he said.) Cain isn't alone. Other detractors would eventually tell me, as long as I didn't disclose their names in print, that BedBug Central is a "shyster organization," that they charge tens of thousands of dollars to endorse a product, and that they've stolen product ideas, sometimes from former business partners.

From another vantage point, the Coopers and BedBug Central could be considered a model of the American Dream: they identified a gap in the market early on, built a relevant business, and made it profitable. The company has encouraged affiliates to donate more than $300,000 worth of bed bug treatments to low-income homes and shelters over the December holidays since its inception, although without mandatory follow-ups it is impossible to track if they've all been successful; and Rick Cooper helped guide New York in the early years of the city's resurgence. The company also has many fans. In an interview, one told me that BedBug Central's approach was a "marriage between university objective research and the private marketplace," and said that he shuddered to think where the industry would be without it.

When I pressed Phil Cooper about his critics, he countered that making money isn't a bad thing. When he started BedBug Central, he said, "I saw all these possible ways of leveraging and truly making money. I mean, let's make no bones about it. BedBug Central is not a nonprofit." But, he added, "when people say BedBug Central is a Walmart and will sell anything that doesn't work, I will tell you that is not true. . . . We're very selective." He also lamented the rumors that bed bug companies have to pay to partner with Bed-Bug Central, admitting that while it's true, it isn't nearly as much money as people have claimed—more like $7,500 per company per year than tens of thousands—and that it is a model he eventually changed because too many partners have complained.

He wants to keep them among his happy fans. Of these, Cooper told me: "The people who have gone through BedBug University boot camp here? If you talk to them and interview them? They will tell you they think we're God."

• • •

Similar dramas play out between emerging bed bug businesses across the globe as companies jostle for attention and market share, although most of it is happening in the United States. (The UK and Australia have stricter regulations on new products, which might account for some of the tension between the Coopers and the Cains of the bed bug world; it's not just a question of ethics, but of culture.) BedBug Central's $1 million annual revenue represents just a drop in an industry worth at least hundreds of times that. In 2011 bed bug businesses in the United States alone earned more than $409 million, and that only included professional products and services such as insecticides, canine detection, heat treatments, and other tactics that generally require an expert touch. Sales of direct-to-consumer products—of which there were more than 8,000 on Amazon and around 13,000 on Google Shopping by 2013—aren't included in this figure. Neither are lawyer fees, the cost for rebuilding after accidental fires or recovering from unintended insecticide poisonings, nor new commercial insurance policies, which protect pest control operators or cover bed bug damage for hotels, landlords, colleges, and companies who send their employees on business trips.

Even the bed bug itself has become a commodity as researchers and canine units have sought live bugs to start up colonies in the lab or to train dogs. An intrepid entomology company in England called CimexStore sold between 35,000 and 40,000 live bed bugs in 2012 for prices ranging from 20 pence for a first-stage nymph to £1 for an adult. The company also sells dead bugs from each life cycle preserved in resin as a demonstrative tool, as well as custom display cases with bed bugs arranged in a mock mattress corner. In the United States, while some entomologists price their bugs at around $1 or $1.25 apiece to cover rearing costs, private companies may charge between $2 and $6. A more expensive and high-tech option for dog trainers is to purchase CimexScent, which are strips of paper steeped with the odor of live bed bugs. These cost $120 for a five-pack, including various shipping costs.

In 2011 the bed bug market was growing. While the bugs were only the sixth highest-grossing pest in the United States—ranking after ants, termites, cockroaches, rodents, occasional invaders such as silverfish, and spiders—they were the fastest-growing moneymaker in the industry. By 2012 bed bug revenue dipped to $401 million, although business surged in the western and midwestern states compared to the South, and bed bug cases were up overall. Worldwide, too, the bugs continued to be a problem. In an industry survey published in 2011, pest control operators from Africa, Australia, Europe, and North America reported that bed bugs were the most difficult insect pest to control, more so than ants, termites, or the formidable cockroach, which lore claims can withstand a nuclear blast. In the United States, bed bug treatments increased ninefold in the past decade.

To get a sense of what the future might hold and whether companies would continue to capitalize on bed bugs, I trawled the US Patent and Trademark Office databases for hints of upcoming products. Soon I had built vast spreadsheets and graphed all existing bed bug patents over time. As of 2012, nearly five thousand patents and published patent applications that related in some way to bed bugs had been filed worldwide. Most were released in the last forty years, although those from the seventies through the nineties were insecticides that broadly claimed bed bugs along with hundreds of other potential insect targets so the patent owners could cover all of their chemicals' possible uses.

The biggest boost happened around the last decade: between 1992 and 2012, published patents and patent applications jumped by 960 percent. Most were insecticides, although few would ever make it through the vigorous efficacy and safety testing requirements to become a product, let alone one for bed bugs. (This was true even when Paul Müller and J. R. Geigy began searching for what would eventually become DDT in 1935, as Müller once lamented: "The situation looked desperate indeed. Already an immense amount of literature existed on the subject and a flood of patents had been taken out. Yet of the many patented pesticides there were practically none on the market. . . .")

Between 1992 and 2012, there was also an increase in patents for heat and freezing systems, mattress encasements, bed-bug-resistant furniture, repellents, attractants, and biological agents. The second-largest category after insecticides, though, included surveillance tools such as traps, monitors, detectors, and barriers—the same bed bug technology first patented more than 150 years ago. The newer versions included electronic noses with chemical sensors that allegedly sniff out bugs better than a dog. One version exploits a species of stingless wasp trained to associate the smell of a bed bug with sugar water, which the wasps like to eat. The wasps are held in a canister, and cameras inside it chart their movements. According to the inventors, when the wasps smell a bed bug, they grow excited; the camera notices their frenzied movement and triggers an alert. Future traps may also pull inspiration from the meat-hook ability of bean leaves, as the California scientists who discovered this feature are working to license a synthetic version to sell as mats, clothing, or strips to wrap around bed legs or headboards. Or there may be pheromone-laced traps, which the researchers at the London School of Hygiene and Tropical Medicine are trying to commercialize.

The products for other recent patents are already on the market. There are the various takes on the plastic bed-leg saucer traps, of course, as well as boxes with carbon dioxide or pheromone decoys and a wide range of sticky traps, although scientists debate whether glue-like surfaces even work. But that doesn't stop the sales. Take BuggyBeds, a glue trap laced with various chemicals thought to attract bed bugs either to humans or to other bed bugs.

US 20120324781A1

(19) **United States**
(12) **Patent Application Publication** (10) Pub. No.: **US 2012/0324781 A1**
Smiley (43) **Pub. Date:** **Dec. 27, 2012**

(54) BED BUG AND ROACH TRAP

(76) Inventor: Everett J. Smiley, Oakford, IN (US)

(21) Appl. No.: 13/135,097

(22) Filed: **Jun. 24, 2011**

Publication Classification

(51) Int. Cl.
A01M 1/10 (2006.01)

(52) U.S. Cl. 43/123

(57) **ABSTRACT**

The present invention relates generally to traps for bed bugs and roaches. Bed bug infestations have taken a dramatic upturn in the last several years and roaches are among the oldest surviving creatures on earth. This trap takes a novel approach to luring bed bugs and/or roaches out of their hiding places, using a live warm blooded animal to entice them into the trap and killing them once they enter the trap by cutting their legs off with a rotating cutter blade, shearing off any protrusion through the perforations in an interior wall.

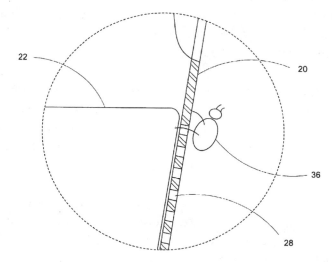

A published patent application for a bed bug trap that uses a live warm-blooded animal as bait and then cuts the bugs' legs off with a rotating blade. Credit: United States Patent and Trademark Office.

The trap featured on the 2012 season premiere of ABC's reality show *Shark Tank*, which drew a reported 6.4 million viewers. In the show, entrepreneurs pitch businesses to a panel of five angel investors (the Sharks) including, most famously, Mark Cuban, a billionaire entertainment mogul and the owner of the Dallas Mavericks.

In the episode, after a business that sold equipment for masseuses who use only their feet failed to interest the Sharks, all five perked up when two women from New Jersey, Maria Curcio and Veronica Perlongo, came in with a faux bedroom set and Curcio announced: "A simple twin bed like this can have thousands to hundreds of thousands of bed bugs feeding on you, causing you red, swollen, itchy lumps!"

The Sharks, each dressed in a sharp suit, scratched and fidgeted when Perlongo added: "You could have bed bugs in your home and not even know it!" The Sharks leaned forward when Curcio said the product would "attract, trap bed bugs dead!" But they got really excited when the women said they already had $150,000 in sales through stores such as Burlington Coat Factory and Home Depot and that they had turned down an offer of $5 million for their patents and trademarks because they thought BuggyBeds was worth more. All the women were looking for was help breaking into box stores such as Walmart. For the first time in the history of the show, all five Sharks made a joint offer ("I'm itching to do this deal!"), and the New Jersey ladies accepted.

Within weeks, the agreement was no longer in place, although neither BuggyBeds nor ABC would tell me why when I tried to arrange for interviews. Still, the product was soon available at Shop-Rite, Associated supermarket, Met Foods, True Value Hardware, Ace Hardware, and on Amazon. A disclaimer on the BuggyBeds website reads: "BuggyBeds detectors are not intended to prevent, destroy, repel, or mitigate, a pest. Their sole intention is only to attract pests for detection purposes."

• • •

The problem with professional bed bug businesses, an industry consultant once told me, is the disconnection between a customer's expectations and reality. Bed bugs are disturbing for people in a way that other pests are not, so the assumption is that professionals will be able to get rid of the bugs. This is the desire with other pests, too, but the tolerance for a roach or two is, perhaps, higher than for a bug that bites. It's called the pest *control* industry and not the pest *eradication* industry, the consultant pointed out. A pest eradication industry would make for a bad business model; if the exterminators got rid of every single creepy-crawly in all of our homes, they'd eventually go out of business. Instead, exterminators need to keep some pests around, the consultant told me, choking up for a moment because of this injustice. Later, when I posed the issue of control versus eradication to other bed bug companies, including those that the consultant said were the worst, I was told that they were strictly eradication-oriented and that it was other companies—bad companies—that would settle for mere control. And, of course, the consultant had his own list of recommended vendors and services to sell.

Still, it is true that completely wiping away any pest even from a single dwelling can be arduous, particularly in dense locales where people live pressed against one another. In the city, my rats are your rats, as are my roaches. And my bed bugs? We share those, too. The pests work their way through abutting walls or down shared hallways or across joined basements, unaware of the invisible legal boundaries determined by leases and deeds. Even stand-alone houses in the suburbs can be hard to treat, with larger expanses that come at high costs, although they have a smaller chance of being infested again by a neighbor. And in cases where a bed bug is expunged from a specific apartment or house or building through a meticulous and pricey treatment, all it takes are bed bugs to move in from next door or a single fecund female hidden in a traveler's suitcase to start a new infestation.

Expectation versus reality is also a problem with over-the-counter bed bug products. We want to believe the promises on the labels, especially in the moments of insomnia between two a.m. and dawn when we lie awake in our beds, fully clothed under the sheets and white-knuckling a flashlight, ready to catch our tormenters mid-bite. We want a single miracle product that will protect the sanctuary of the bed, despite the growing piles of useless weight-lifting contraptions and anti-wrinkle creams and vegetable choppers stashed in the dark corners of our closets and junk drawers. Usually, this belief is based on a murky promise. The company may claim that a product "Kills Bed Bugs on Contact!" and the consumer mentally stretches it to mean complete annihilation of all the bed bugs in the home.

The bed bug market explosion and our bed bug amnesia have provided the perfect combination for what economists call asymmetric information, a theoretical case where one party in a transaction is less knowledgeable of the capabilities of a product compared to the other. Without the technical or scientific savvy—specifically, knowledge of the underpinning biology, psychology, or behavior of a pest and the ways a product could exploit these characteristics—the consumer is unable to make an educated decision on what to buy and what to avoid. In a free-market economy, of course, pretty much anyone can make a product and sell it. They are also relatively free to exploit their customer's lack of knowledge on how something actually works or make wild claims about its capabilities, at least until the wrong people, from the seller's perspective, start to pay attention.

• • •

In America, crushed expectations are often followed by lawsuits. Among the most famous in relation to bed bugs is a reported $10 million complaint against the Waldorf Astoria in New York City. The plaintiff said she was bitten by bed bugs during a stay at the hotel in 2007, and that the trauma lingered until she decided to take legal action three years later. As of late 2013, the case was still wending its way through the Supreme Court of the State of New York. The suit was one of several against the luxury hotel, and the Waldorf Astoria is certainly not alone, as other cases have been filed against hotels and motels across the country (most are not for such a large sum of money). While it is usually people staying in the hotel who sue, that is not always the case. In Canada in 2013, the Hotel Quebec filed a complaint against a guest who had written a bad review claiming he got bed bugs during a stay. The press picked up the review, which the hotel claimed was false and had hurt its reputation, leading it to seek a reported $95,000 in damages.

Although it is hard to track all bed bug cases, which mostly occur in state courts, arguably more common than the hotel lawsuits are disputes between tenants and landlords. The largest of these cases was filed in the Iowa Supreme Court in 2010 as the first-ever modern bed bug class-action suit, with around three hundred tenants of a low-income apartment complex suing their landlord after allegedly suffering two years of bed bug infestations without proper treatment. The plaintiffs sought a reported $7.4 million in damages. The highest reported payout to a single tenant comes from Annapolis, Maryland, where a jury awarded a woman $800,000 in a lawsuit against the woman's landlord, whom she said had rented her a bed-bug-infested apartment.

More common are the smaller lawsuits, such as that of a thirty-six-year-old Minnesota woman I'll call Carrie. In the summer of 2012, Carrie arrived at her new apartment, a cramped two-bedroom twenty minutes northeast from Minneapolis, to find that the previous tenant had left most of his belongings behind. The apartment was filthy, too, reeking of cigarette smoke and wet dirty dog; the walls were yellow and the floor matted with hair. She had been forced to rent the apartment after a painful foreclosure on her house in the city, a casualty of the US real estate collapse. The apartment's chaotic mess did not help her despair.

Her new landlord was an acquaintance she'd known for more than a decade through a singles group where they played baseball, barbequed, and had board game nights. She'd needed a place; he'd offered her this. Once she was in the dirty apartment, there was not much she could do. She'd already paid an $1,800 deposit and the first month of rent, and she hadn't bounced back financially from a layoff and subsequent stint on unemployment a couple of years prior to find another rental. The redemption period was practically up at her old home. All she could do was unpack.

Before she could unload the van, she had to clean the apartment. It took a small army of friends most of the day to clear out the old furniture and trash, including hacking apart an oversize smelly couch to fit it through the door. Another six hours went to peeling yellowed layers of tobacco smoke from the walls with bleach and a scrub brush. The carpet was hopeless, but the landlord assured her that he'd take a look at it, so she moved in her new bed set and couch, as well as her three dogs and their beds, and tried to make it their home.

After just a few days in the apartment, Carrie woke one morning with itchy lumps on the inside of her left wrist. By the sixth day, her face, neck, and shoulders were ravaged with red hives, and her left eye was swollen near-shut. She went to a local health clinic and left with a prednisone prescription, thick cream, and no definite diagnosis other than a bad allergic reaction. She packed the dogs and a few things and drove to her parents' house to recover. After a few days, the hives and rash vanished. A friend had suggested she may have bed bugs, and so they returned to the apartment together armed with spotlights and plastic baggies. They found their game quickly: there were several bed bugs under the bed skirt, in the dogs' kennel, and caught in a spider's web in the corner of the bathroom. They bagged them up and left.

Carrie moved out the next day and put all her belongings in a storage unit. She doused her bed with malathion, an organophosphate that hadn't been legally registered for indoor use in the United States for several years, and came back a week later to find around half a dozen dead bed bugs on the mattress. She sprayed again and coated other pieces of furniture with diatomaceous earth. She broke her lease and asked for her deposit and rent money back, citing a portion of Minnesota state code protecting tenants from uninhabitable living situations involving insects and vermin.

While she waited for a response, she went back to talk to her neighbors, with whom she had grown friendly despite her short time in the building. The family directly downstairs from her rental—an unemployed middle-aged couple who lived with their two teenage sons, their twenty-three-year-old daughter, and the daughter's four-year-old—told Carrie that the bed bugs had been a problem in the building for more than a year. In fact, the daughter, her child, and one of the teenagers had been sleeping in a tent in the yard for months to get away from the nightly bites, while the other teen slept in the building's basement on a plastic lounge chair. The parents ritualistically sprayed the couch with alcohol before going to sleep, which came in spurts as they woke up to apply more throughout the night.

Across the hall lived another woman, partially disabled from a stroke and only able to work part-time. She had bed bugs, too, but wouldn't admit it, although the other tenants began to suspect that she was the source of the building's bugs after their landlord made everyone throw their beds in the communal dumpster; hers, a freebie her son and his girlfriend had dragged in a year before, by far had the most bed bug casings and feces. There was a rumor that the landlord, who lived in the building, also had bed bugs, but he wouldn't admit it, either. He also refused to respond to Carrie's request for a refund, so she filed a civil complaint and hired a housing lawyer. They eventually settled for $2,700. Within a month, her landlord had already rented the apartment to a man who paid a year's rent up front in cash.

Although I wouldn't call Carrie lucky, it is fortunate that she lived in a state with bed bug laws. Minnesota is one of just twenty-two states with such laws, and most of them were enacted in response to previous lawsuits and other legal disputes. Congress introduced the Don't Let the Bed Bugs Bite Act of 2009, which was meant to support states in inspecting hotel rooms and public housing for bed bugs, but it was not enacted. And while older laws often specify a landlord's responsibility for keeping buildings "vermin-free," they don't always list bed bugs, which leaves a small window for debate and uncertainty.

Whether any of the laws will deter bed bug lawsuits isn't clear. In the past, the older vermin laws did not. In 1931 a landlord and tenant in St. Paul, Minnesota, fought about a bed bug infestation in court, which the tenant had unsuccessfully treated with twenty gallons of gasoline. A judge ultimately awarded the tenant $50.95 for rent reimbursement and legal costs. And in 1887 a New York court ruled that the tenant in another bed bug case was out of luck, noting that bed bugs and a number of other vermin should be no surprise to anyone renting in the city.

• • •

During the final stretch of presentations on the last day of the Las Vegas summit, two representatives from the Environmental Protection Agency took the ballroom stage. They were there to address regulatory actions that help protect consumers from products that don't work. A few members of the audience shifted in their seats and muttered. In side conversations both before and after the summit, several confided to me that the EPA was slow to acknowledge the bed bug problem or provide any guidance or support. Worse, they said, was the agency's treatment of the slippery category of 25b chemicals, the minimum-risk pesticides. According to the representatives onstage, who had to raise their voices to mask the crowd's murmurs while doing their best to ignore them, the agency was working on the 25b problem. But so far, companies hadn't been cooperative with providing information to inquiries about efficacy data. The audience grumbled as the reps left the stage.

For the next and final talk of the summit, an attorney with the Federal Trade Commission stepped to the stage. The audience quieted down. The FTC protects consumers from false advertising claims and, along with the Department of Justice, enforces antitrust laws. The FTC's most well-known rule is the Do Not Call Registry, which shields people from unwanted telemarketing calls. As the attorney flipped through her introductory slides, I began to realize that, unlike the civil lawsuits that hotels and landlords faced, companies had a different sort of legal action to consider.

The attorney explained that under Section 5 of the FTC Act, which outlines unfair or deceptive acts or practices, any company must be able to provide legitimate and reliable evidence that supports certain advertising claims. In other words, if a bed bug product's advertisements or marketing says it kills bed bugs dead, that company has to have the scientific studies to back it up. Companies also must turn those studies over for review at the commission's request, and the tests have to pass muster with third-party experts. The entomologists in the crowd started murmuring again, but the energy had shifted from grumbles to an animated buzz. It wasn't clear how many vendors were present, since many had packed up or already left, or whether they'd find the news exciting or sobering.

Scattered applause broke out as the FTC attorney pulled up a slide entitled "Recent Law Enforcement Actions," which listed the agency's first legal responses against two companies marketing 25b bed bug products. One was Rest Easy ("Kills & Repels Bed Bugs, For Organic Use," according to the label), made with cinnamon, lemongrass, cloves, and peppermint. The other was Best Yet ("Chemical Free Insect Control" for bed bugs and other pests), with cedar oil as the primary ingredient.

The mood dampened when the attorney explained that the sprays could remain on the market (and indeed, months later, they were still available at stores and online). Depending on the case, the FTC only has the power to force changes in advertising claims, to pass out financial penalties, or to require restitution for people who were tricked into buying a bad product. The fines could ruin a company if the penalties are high enough or if a large number of customers are entitled to refunds. Repeat offenders could even be barred from ever participating in the bed bug market again. It wasn't perfect, but it was a start.

EIGHT

MYSTERIOUS RASH

The Psychological Toll of Travel

I'll never get used to that, I thought to myself when I walked into David Cain's office to find him feeding a small jar of bed bugs on his forearm. He was dressed in his customary black leather pants and black golf shirt, the same uniform I'd seen him in at the BedBug University Summit in Las Vegas eight months earlier and at our follow-up interview in Queens nearly two months after that. Cain's business, Bed Bugs Limited, was tucked away in a South London industrial park, and his small office smelled of stale cigarette smoke and was crowded with papers, boxes, and a little glass dome full of rice left over from lunch. Later I learned that the rice was for his pet Madagascar hissing cockroaches hiding somewhere under the cardboard that lined the bottom of the enclosure. Cain held the jar of bed bugs in place by squeezing it between his left arm and his chest, leaving his right hand free to work on the computer that sat on his desk amid stacks of paper, mugs, and an ashtray. As he typed, he lectured about bed bug academics with their "hands in the cookie jar" and "fingers in the pie," making money off of bed bug technologies without appropriate disclosures. I asked if I could take a photo. He nodded, and I snapped a couple with my phone.

It was my last day of a four-day layover in England, a short stop on my way to the Czech Republic. After searching for possible threads to follow, I had decided to explore in person the idea that the bed bug resurgence had originated in Eastern Europe, as well as to continue Robert Usinger's unfulfilled adventures in the region, if only in spirit. But before that, during my short stay in London, I had hoped to shadow Cain on a bed bug job. He claims a high success rate in his 20,000-plus cases, and at the time of my visit he had recently started a new heat regimen in hotels across the city, which I was curious to see firsthand.

"The reality is, we kind of gave up using chemicals eighteen months ago," he told me after railing against the chemical companies and lamenting that the academic research on pesticide resistance, while interesting on an intellectual level, wouldn't immediately help the pest control industry. "We haven't used chemicals in a hotel room for at least the last twelve."

I asked how the heating system worked, but he wouldn't say. He was also tight-lipped about the rest of his bed bug protocol, as he had been in the past, telling me I would have to sign a non-disclosure agreement to get a look. He would only divulge that his methods work so well that one competitor had come to his office under the false pretense of collaborating, but instead swabbed the furniture looking for chemical traces to figure out Cain's success. Another adversary made knock-offs of Cain's patented bed bug monitor, which were based off a sample that Cain had designed to repel the bugs rather than attract them so that any attempt at ripping off the real design wouldn't work. As for me tagging along on a job, his answer was a flat no. He told me I was not properly trained and I would just get in his way. He was also protective of his clients and their privacy, wary of exposing them to a nosy reporter.

I gave up, content instead to see the bed bug museum that I had heard so much about during our first interaction in Vegas. Rather than a museum in the traditional sense, Cain's collection was stored in four plastic bins filled with various bed bug knickknacks. While Cain had been collecting bed bug ephemera, toys, and books for more than seven years, he had not yet organized them. He pulled out two of the bins—the rest were in storage and unavailable for the day—and we began sifting through the jumbled contents.

The first bin held old pest control advertisements and manuals, as well as public health treatises many decades old. There were photos of Bed Bug Island off the coast of New Jersey, which you can still find on Google Maps, and Bed Bug Alley in New York, which you can't. I picked up an orange and white program for Vladimir Mayakovsky's play *The Bedbug* from a staging at the Mermaid Theatre in 1962. It was the last time the play ran in the United King-

David Cain feeding a jar of bed bugs on his arm. Credit: Brooke Borel.

dom, according to Cain, who seemed surprised when I told him I had seen it in Brooklyn in 2012; I was equally surprised that no one in London had tried to put the show on during that city's own bed bug resurgence. We sifted through nods to old poisons, including a reproduction of a Valmor Bed Bug Murder label, a miniature replica of a pesticide container intended to hang on a necklace, and a glass bottle that once held Bed-Bug Killer Poison, which Cain told me was a brand of DDT. The label read: "Being a VIOLENT POISON it must be kept out of the reach of children, and not where food is kept." A BedBugs board game caught my eye. I lifted it from the bin and looked at the cover photo, which showed disembodied hands using oversize tweezers to pluck brightly colored plastic bed bugs off of a shaking motorized bed. It made me nostalgic for childhood, as well as for the time when I didn't know what bed bugs were.

In the second bin, there were bed bug toys; a bottle of nail polish from the brand Bug Me Not in a color labeled Bed Bug, a light blue pearl; and stacks of antique postcards, which included a series of saucy Victorian images with scantily clad women engaged in bug hunts. A copy of *Statutes at Large of England and of Great-Britain*, volume II, which was published in the early 1800s, had flecks of bed bug markings along its edges and joints; Cain told me it was the oldest book he had found with fecal evidence of a bed bug infestation.

Cain picked up another book entitled *Critical Cities* and thumbed to a chapter on how pest infestations and social reform shaped London in the early 1900s. A black-and-white photograph showed the rubble from slums in Somers Town, a neighborhood that still exists on the city's north side. The slums were destroyed in order to modernize the public housing. In the photo, men in flat caps and overcoats burned effigies of the four pests that plagued them: a rat, a cockroach, a flea, and a bed bug. The men were smiling.

When we reached the end of the bed bug museum, my hour of allotted time with Cain had run out. I left his office and made my way down deserted industrial streets to the Vauxhall Tube station and then hopped on a train toward Kings Cross. I was staying in a crowded hostel there, coincidentally just a short walk from the former Somers Town slums where the bed bug had been burned in effigy more than eighty years prior.

On the train, I squeezed into a seat. Too tired to read, I flipped through the photos I had taken during the interview. I landed on the portrait of Cain in his desk chair feeding the jar of bed bugs on his arm. My side started to itch.

• • •

Two days earlier in an airless plastic capsule masquerading as a shower, I had noticed a rash on my right side. It was red and slightly

Men at the Somers Town slums burning effigies of a bed bug and other pests in 1930.
Credit: Camden Local Studies and Archives Centre.

swollen, a mottled itch in a straight line just above where the waist of my jeans would have hit. Or, maybe, where the junction of my T-shirt and pajama pants could have parted during a restless sleep. I twisted to the right and craned my neck, trying to get a better look. Were there welts?

I had picked the Kings Cross hostel not for its ambience—which, like its shower capsules, was dismal—but for the proximity to the London School of Hygiene and Tropical Medicine and the St. Pancras International railway station. The latter had a regular train to Sheffield, roughly 160 miles northwest of London, home of the University of Sheffield. I would visit bed bug experts at both schools before catching a flight to Prague to continue my adventure. Finding bed bugs in the hostel wouldn't have surprised me, considering the young international clientele who had come to London to tour and to party. There were more than a hundred guests, and while those I interacted with were nice enough, many were messy and left their belongings strewn about the congested dorm rooms.

I thought I had been desensitized to bed bug paranoia after two years of solid research, but as I inspected the rash, I rummaged through my mental collection of images from my previous bed bug bites. The memories ranged from five to nearly ten years old, but even so, this rash didn't look quite the same. There were no distinct welts like I had before. But yet the redness and swelling was familiar, and the rash appeared in the straight line that is common in clusters of bed bug bites. As I toweled off and struggled back into my jeans in the muggy capsule, I thought of the five bunk beds in the room I was staying in, piled with blankets and pillows that guests were not required to bring down to the front desk with their linens during checkout. I thought about the suitcases and backpacks stuffed under each bed, wadded clothes and guidebooks peeking through open zippers. Worse, I thought about the fact that I had slept with my most valuable belongings in a bag on my bunk because the room had no lockers. A lost computer or phone would have made work impossible. I thought about the hundreds of people and suitcases on my red-eye from New York, my train ride from Heathrow Airport to Paddington Station, and subsequent Tube trips. I thought about the day I had arrived when I had shoved my suitcase into a storage room stacked floor-to-ceiling with bags from all corners of the world as I waited for a 2 p.m. check-in.

Back in the room, I dug through my borrowed duvet and its cover and then ran my hands around the mattress, flipping down the seams as I traced along its edges. I found nothing. Maybe they weren't bites, though a glance at the rest of the beds and bulging suitcases fed my doubts. There was no time to check the entire room. I also had no desire to dig through someone else's sweaty sheets or belongings. Even if I did, what if I got caught? It'd be difficult to justify. (*What do you think you're doing? Oh, me? Just looking for bed bugs. Definitely not robbing you.*)

Resigned to the fact that I couldn't solve this mystery yet, if at all, I checked my canvas tote to make sure I had everything I'd need for meetings that day: laptop, recorder, camera, chargers, extra batteries, voltage converter, notebooks, water bottle, wallet, and papers. When I swung the full bag into its accustomed spot on my right shoulder, I noticed how heavy it was. As I walked down a narrow hallway lined with doors to other dorm rooms, the bag rubbed uncomfortably right at the spot of my rash. A second idea began to take shape. Maybe the rash wasn't a line of bed bug bites. Maybe, during hours of walking the previous day, the bag had rubbed my woolen shawl against my side and I hadn't noticed. Maybe the rash was just a rash.

I mulled over my two hypotheses as I dodged morning commuters on a twenty-minute walk from the hostel to the London School of Hygiene and Tropical Medicine, where I would meet with chemical ecologist James Logan to see how his team's pheromone-baited bed bug trap was coming along. With each tidy café that I passed, my position on the rash's origins switched.

By the time I reached my destination, I was debating the position that the rash was from my bag when the school's classic stone façade pushed the argument from my mind. I took in the building's second-floor balconies, decorated with eight unique medically important pests, or at least those considered important when the building was finished in 1929—bed bugs, fleas, flies, lice, mosquitoes, rats, snakes, and ticks—and craned my neck to see the building's famous frieze, which spelled the names of twenty-three scientists who had made some contribution to public health or tropical medicine. Somewhere up there was the name of the school's founder, Patrick Manson, the same Scottish doctor who first discovered that insects are vectors for disease when he found filarial larvae in a mosquito in 1877.

I pushed through the heavy doors and gave my name to a security guard. As I waited for Logan to retrieve me, I entertained myself by reading the glass-enclosed displays of old medical tools and other curiosities that lined the foyer. Among them was a 1946 pamphlet titled *You and Your Children*, which included chapters "Fear of the Dark" and "What Your Children Think about You"; a set of five antique obstetric forceps arranged in a leather roll case, which, disturbingly, reminded me of the tool bags I'd seen farriers use to hold their horse-shoeing gear; and a pair of thick scissors labeled as a Denman's perforator, which, according to the small printed card next to it, was used to puncture a baby's skull to make it easier for it to come out during a difficult birth, before cesareans were considered safe. I shuddered and took a seat.

Brass bed bug detail on a balcony of the London School of Hygiene and Tropical Medicine. Credit: London School of Hygiene and Tropical Medicine.

After about fifteen minutes, Logan bounded up and shook my hand as he apologized for running late. Young and lanky, he was dressed in neat tan pants and a plaid shirt. He ushered me down a sweeping staircase and into a bright basement cafeteria humming with students. We settled into a pair of low couches in a corner and were soon joined by Logan's colleague, an entomologist named Mary Cameron. After some initial chatting about my project and which other experts I'd visit in England ("There are a couple of ones you want to look out for," Logan warned, adding that not everyone knew as much as they claimed, though he wouldn't disclose who on record), the two researchers then gave me a crash course on how their group was unraveling and exploiting the ways in which insects use chemicals to communicate. For their bed bug trap, the scientists described collecting sample after sample of bed bug scat, convinced it contains the aggregating pheromones that help guide the insects home after a tasty blood meal. The scientists were trying to narrow down which chemicals would make the best lure to attract bed bugs.

When time ran out on our formal interview, Logan jumped up and led me to the labs to see the work in action. I chased after his long strides down mazelike halls and stairwells to one of the school's low-ceilinged insectaries. Netted cubes lined the shelves of each vaulted room holding mosquitoes and biting flies, many of which had been collected decades before from far-flung locales. The bed bugs were inside an incubator that looked like a small white refrigerator, but we couldn't peek in because it was too late in the morning, and, like most lab-raised bed bugs, they were on a reversed light cycle to align with their typical circadian rhythm. But a research assistant had one jar of the bugs out, which she had been feeding on her arm to see if she had a reaction. The rash on my side began to itch, and I had another flash of doubt about its origins.

I pushed the thought from my mind as Logan ushered me into a claustrophobic white-tiled chamber. The room was nearly empty, save for a single cot and a package of silver thermal bags intended for an outdoor survival or trauma kit. Here in the lab, the bags were used to collect the volatile chemicals that emanate from a human body called kairomones, which mosquitoes and other insects, possibly even bed bugs, use to help locate a meal. In an experiment, a person would get inside the bag, and plastic tubes would pump filtered air in and pull the mixture of the air and the person's body odor out. The scientists could capture the odors and feed them to a device called a mass spectrometer to help characterize their chemical composition (they used a similar, though smaller-scale, technique to collect pheromones from bed bug feces). Then the scientists could take bed bug heads—the cells of which remain active for about thirty minutes or an hour after decapitation—and connect the antennae to delicate electrodes to measure the reaction to various kairomones or pheromones. Stimulations that matched the timing of a particular scent showed that bed bugs could, indeed, detect the chemicals.

In another lab, sealed off from the outside light like a photography darkroom with a heavy black curtain and lit with an eerie red light, I watched as another of Logan's colleagues, a young medical entomologist named Ailie Robinson, showed me how to test live bed bugs on blends of chemicals that may work for a trap. A small light box with a clear plastic dish would soon hold a bed bug, and an internal set of tubing would feed a test blend and a control through two distinct openings. Motion-capture software would track the bugs' movement to show whether they clustered around the proposed trap scent or the control, or if they just wandered aimlessly. Robinson handed me a small plastic jar that held the next set of bed bugs to be tested, and the researchers encouraged me to take a whiff. I held it under my nose and breathed in the cloying musky smell. I thought about my rash.

Logan guided me back out of the maze of labs and, late for meetings and a television appearance, handed me brochures on the school's commercial consultancy division, of which he is the director, and bid me good-bye.

• • •

The next morning, after another sleepless night of spooning my luggage and wondering whether computer theft or a bag of bed bugs was worse, I caught the 7:55 a.m. train from St. Pancras to Sheffield. I spent the two-hour ride going through my notes, dozing, and thinking about my overnight backpack squished between suitcases on the train's luggage rack, as well as my larger suitcase back in the storage room at the Kings Cross hostel. After my train pulled into the station at Sheffield, I darted through a late morning drizzle and flagged down a cab toward the university. I slammed the door, and as the driver pulled away from the curb, I stared out at the rain as it hit the windows and wondered how many passengers had sat in my seat on that day. That week. That month. My side itched.

When I arrived at the school, I walked past a bored security guard and rode an elevator to the B2-level floors to the Evolutionary and Ecological Entomology research unit to meet Michael Siva-Jothy, the same scientist I had seen introduced to the stage with LMFAO's "Party Rock Anthem" at the BedBug Summit in Las Vegas. Siva-Jothy was again dressed in a neat suit, this time in preparation for a policy briefing at the Royal Society in London later that day. We sat at a long table in his sparse corner office. The screen savers on three Mac computer monitors on his desk flipped through gorgeous macro photographs of insects—a fly with compound eyes like wraparound sunglasses, a ladybug captured in flight, a pair of bed bugs in the midst of their traumatic mating ritual.

Siva-Jothy gave me a rundown of the bed bug work in his lab, which was going through an overhaul after several key researchers had either finished their degrees or otherwise moved on to career advancements at other institutions. After about twenty minutes of this, I nudged the conversation toward his bug research in Kenya, which he had brushed off at the BedBug Summit. He had already warned me that my chances of joining his next African expedition were slim, even if the funding for the trip did come through. For the moment, I would have to be satisfied with living vicariously through his past adventures.

In 2004, armed with Robert Usinger's *Monograph of Cimicidae*, Siva-Jothy and two colleagues—his master's student Richard Naylor and a tall and angular German entomologist named Klaus Reinhardt—flew to Kenya in the hopes of finding a variety of bugs. The region provided the broadest possible range of interesting species in the most manageable geographical space. One was a little-known African bat bug called *Afrocimex constrictus*, a close relative to the common bed bug but with an even more complicated reproductive system, which is why the entomologists were interested in studying it. Usinger's book described the caves where he had found the same species fifty years prior, and his work was the only published account of where many of the caves with the bugs were located.

"We knew if you were going to try and find a particular species of bed bug or bat bug alive, Usinger's notes were so compelling and complete—and actually pretty accurate—they were the best jumping-off point for identifying where to go," Siva-Jothy explained, as he launched into his story.

The trio, along with two African guides named Joseph and George, spent three weeks trekking the length and breadth of Kenya in a minibus looking for Usinger's caves using photocopied pages of his monograph because Siva-Jothy only owned a first edition of the book and it was too rare and valuable to risk the trip. Along with Usinger's directions, the team asked locals if they knew of any nearby caves that had bats. Where there were bats, there may be bat bugs.

The scientists found many caves. At least two dozen. In addition to Usinger's suggestions, there was a cave at a diatomaceous earth mine near Nairobi that allegedly had a fully mature bat colony, but the miners turned the scientists away out of fear of liability. There was the cave in Kiambu, also outside of Nairobi, in Paradise Lost Park, where the scientists saw bronze sunbirds, cinnamon-chested bee-eaters, and black-backed puffbacks. The cave, however, turned out to be so well toured that it had electric lighting and was devoid of the wildlife they sought. There were the Shimoni caves in Mombasa on Kenya's Indian Ocean coast, which formerly held slaves en route to a Zanzibar market and are now a tourist attraction. These had no bats, and thus no bugs. And there were the Smelani caves, which were only accessible when the tide was out. These the entomologists found with the help of a local self-described holy man. The caves, he said, were sacred; he would lead the team there, but they could only enter if they paid five Kenyan silvers to a shrine, prayed for something good to happen, and removed their shoes. The scientists complied. The cave floors, Siva-Jothy told me, were "heaving with cockroaches" that numbered in the thousands, and he stepped on a thorn covered in bat guano, which embedded into the ball of his foot. All for nothing. There were thousands of bats, but no bugs.

The team's luck changed at Mount Elgon, an extinct volcano towering over 14,000 feet on the border of Kenya and Uganda. It is the centerpiece of Mount Elgon National Park, a reserve sprawling more than 277,000 acres that are home to leopards, water buffalo, blue monkeys, colobus monkeys with their skunk-like black-and-white coats, hyraxes, and more than three hundred species of birds. The area has elephants, too, Siva-Jothy said, and herds wind through the many caves that dot Mount Elgon's slopes, scrapping their tusks on the cave walls to unearth salts that are key to their diet. At this fact, I stopped Siva-Jothy's description. The elephant story sounded similar to one I had read in Richard Preston's *The*

Hot Zone, which describes outbreaks of the deadly hemorrhagic fever caused by the Marburg and Ebola viruses in the eighties and nineties.

"It's those caves," he replied.

"Those same caves you're going to?" I asked.

He nodded, and it occurred to me that, despite his calm and polished demeanor, I might be dealing with a madman. Perhaps it was better that I would continue my own bed bug adventure in the Czech Republic.

During the 2004 trip, the first of three, Siva-Jothy confessed that his team was ill prepared for a cave that may harbor Marburg virus. They armed themselves only with paper spray-paint masks, protective suits, gloves that could withstand a rabies-tainted bat bite, and headlamps when they first explored Ngwarisha, Makingeny, and Kitum, the latter the same cave made famous by *The Hot Zone*. Indeed, when they first stepped into Kitum cave, they hadn't bothered to put on even this minimal protective gear. In that cave, the German entomologist Reinhardt shouted, "Oh my god, oh my god!" after finding three live bat bugs crawling on the guano-covered floor, his cries startling hundreds of bats into flight. Upon further inspection, the scientists found more bugs in the cracks in Kitum's walls and many more in nearby Makingeny cave, which was teeming. After breaking for lunch and finally putting on their safety gear, the scientists went back in. Eventually, they would identify the bugs as *Afrocimex constrictus*, the very species they sought.

In the coming days and on subsequent trips over the next several years—when they arrived more prepared with full respirators—the team fell into a routine. When the sky grew light at six o'clock each morning at their campsite, a mere clearing in the woods at the foot of Mount Elgon, they would wake to the smell of eggs, toast, and coffee warmed directly on an open-flame fire. After a rain shower, which started each morning like clockwork and lasted half an hour, the scientists would climb into the minibus and drive forty minutes to the volcano. From there, they hiked through a podocarp forest with a bamboo understory, past red-fronted parrots and colobus monkeys, all the while fighting altitude sickness. At the crest, the bamboo would give way to volcanic rock and tuft

Pages from Michael Siva-Jothy's field notebook, Kenya 2004. Credit: Michael Siva-Jothy.

grass all the way to the mouth of the caves. After a proper cup of tea, they would suit up and put on fanny packs, which held the soft tweezers and plastic ethanol-filled tubes necessary for collecting the bugs. Then they would go inside.

"Have you ever been in any bat caves at all?" Siva-Jothy paused to ask me.

"No," I admitted reluctantly, feeling poorly traveled.

"They're all pretty much the same. Pretty smelly, 'cause you're up to your knees in bat shit." The Mount Elgon caves, he added, were also treacherous and "very, very scary." The temperature soared more than 80 degrees Fahrenheit, mostly from the heat radiating from hundreds of thousands of Egyptian fruit bats that roosted overhead. Agitated when people enter the cave, the bats would squeak in hundreds of thousands of simultaneous vocalizations and flap their wings, a ripple of alarm whipping through the colony. As the bats moved, and as the attractive bright lights of scientists' headlamps cut through the darkness, louse flies, bat flies, fleas, and bugs fell from the ceiling or dive-bombed like kamikaze pilots. It was "like walking in a gentle rain of things hitting you," Siva-Jothy said. "If you've got another ten minutes, I'll tell you everything that wants to eat you in the cave. And they go straight for your warm bits."

There was also the unsteady footing, made from piles of rubble that can rise up to five feet high and fall into glacier-like crevices. This precarious landscape changed often, each time the ceilings, crisscrossed with deep cracks, caved in. There was no guarantee that this wouldn't happen during a bug-collecting mission. There were also the elephants, which could barge in at any moment to get a salt fix. And then there was also the threat, though small, of insurgents sneaking over the Uganda-Kenya border, a possibility that required the presence of park rangers armed with AK-47s. But it wasn't these dangers that kept the African guides from venturing in. Neither Joseph, a Maasai who carried a club to fight off aggressive hyenas, nor George, whose front incisors had been knocked out at puberty to signify his manliness, would enter the caves because they were filled with bad spirits, from Siva-Jothy's recollection.

After about two and a half hours in the caves, which the scientists spent collecting bugs and gluing data loggers to the cave walls to measure temperature and humidity, they would stumble back into the sunlight and strip off their protective suits and fanny packs, dumping the whole lot into large trash bags. Next the team would inspect themselves and one another for ticks and biting insects. Once clear, they would hike back down the volcano, through the tuft grass and podocarp forest, to the minibus. Back at camp, after bathing in a spring-fed corrugated shower and having a dinner of roasted goat, they would crawl into their tents not long after the six o'clock sunset. A few times during the night, Siva-Jothy said, he would wake with a start, turn on his light, and then search his body for what he was certain were biting insects from the caves. This, despite the fact that back at Sheffield, he regularly dropped his trousers to feed bed bugs on his bare legs. In the tent, he wouldn't find a thing.

"Why risk all that to get some bugs?" I asked, as our interview came to an end. "Why go into such a potentially dangerous place?"

"Because the science excites me. Why do a job that doesn't pay me enough? Because I'm a kid who gets a massive hit from discovering stuff," he replied with a shrug. "If you don't take the risk, you don't get the science. I think it's as simple as that. That would be one explanation. The other is we're all going to die anyway. Frankly, I would much rather die falling down a crevice in a bat cave rather than getting hit by a bloody bus while I'm cycling in to work."

• • •

After spending the rest of the afternoon touring Siva-Jothy's bed bug lab and watching a graduate student dissect bed bugs under a microscope, I left the university and headed toward the hostel I had booked for the night. The trudge was all the worse for the steady rain that fell from a steely sky. The streets were slippery and seemed to angle in directions that weren't in my favor. Finally, after twenty-five minutes and two wrong turns, I scrambled up one last steep residential lane. At the top, where the street hit a dead end, was a quaint white stucco house. I made my way to the front door through a front lawn overgrown with dandelions. To one side was a trash pile with three mattresses stacked beside it. *Hmmm*, I thought to myself, eyeing the beds, which appeared new and unstained. Once again, my side began to itch.

Inside, a pair of young Spanish men named Sergio and Pablo checked me in, one working an ancient computer while the other hovered nearby. Both were unemployed engineers who had come to England looking for work, and they were helping out at the hostel in exchange for a free place to stay. I asked them about the mattresses, and they shrugged. I asked them if they knew what bed bugs were.

"Bad bugs?" Sergio asked.

"No, not bad," I said. "Bed. Bed bugs."

"I don't know these."

My room was the only all-women dorm, with four single beds side by side. Each was covered with a musty polka-dot duvet and brown sheets; the floor had a patterned carpet that looked like it hadn't been vacuumed in a good long while. As with the London hostel, the only options for storing my bag were the floor or the bed. My roommate—a young woman from Shanghai who would later reveal that she was traveling solo through England on a self-guided Jane Austen tour—had left her suitcase on top of her bed. I sighed and did the same. As I dropped my bag, a tiny dark form skittered next to it and I jumped. I stared at the small speck, slowly coming closer. The thought of checking out and finding another place to stay in the rain was unappealing, but there were no other rooms I could swap to at this hostel, and I wasn't about to willingly sleep with more bed bugs. Once my face was inches from the mattress, I realized the dark fleck was nothing but a piece of black lint.

The walls of the dorm felt like they were closing in, so I left the room and walked down a creaky set of carpeted stairs, which could also use a good sweep, and explored the first floor. In the damp kitchen I found a washing machine with a large sign taped to it that read: "Out of Order." Outside the window, a clothesline dripped in the rain. *How are they washing the sheets?* I wondered. Even if my rash really was from bed bugs and I wanted to wash my clothes in hot water or toss them in a dryer, it wasn't an option. I glanced around and, seeing that I was alone, I peeked at my side. It was no longer swollen, and there was still no definite evidence of bites.

My phone buzzed. It was a message from Klaus Reinhardt, Siva-Jothy's German colleague, who lived in Sheffield part-time. We had plans for a brief catch-up, and I soon found myself down the street in a brightly lit pub, talking about bed bugs with him over a couple of pints of beer. Reinhardt had recently returned from a trip to Austin, Texas, where he had been searching for *Primicimex cavernis*, thought to be the most primitive example in the family of bed bugs, Cimicidae. He and an American entomologist had searched for *Primicimex* in a hot cave that was part of the state's formerly massive bat guano mining industry (the nitrogen in the guano makes excellent fertilizer). Although the researchers wore respirators to protect against clouds of poisonous ammonia and, allegedly, airborne rabies that could be spread from the bats, he still got a lungful of the cave's harsh air, which he described, with a grimace, as terrible. They did not find *Primicimex*. After a short conversation about Reinhardt's own forthcoming book on bed bugs, the details of which he was cagey about, we finished our beers and said good-bye.

I returned to the hostel to find around twelve rowdy men in their late thirties or forties crowded in the kitchen drinking brandy and beer and watching a soccer match on a laptop. Unlike the young party seekers at the hostel in London, I would learn that these men were day laborers trying to get work in a bad economy. They came from all over England, as well as from Lithuania, Russia, South Africa, Scotland, and Spain. Most had jobs installing solar panels with a local energy firm. There was also a freelance chef, who cornered me for a half hour with a sad story about his estranged family, and a former finance guy who was in the midst of finding his purpose in life, who showed me the best place to find fish and chips.

Someone handed me a small glass of brandy. I sipped it, which drew a laugh from a guy who spoke little English, but who showed me how he preferred his brandy by drinking shots in quick succession. I tried talking to a few of the men about bed bugs, but most, like Sergio the Spanish engineer, did not know what I meant. After a couple of hours and a few refills, impossible to refuse as the bottle always tipped my way as soon as I'd finished a glass, I snuck upstairs, turned the key in the door, and fell asleep, protected against bed bug worries by a blanket of brandy.

• • •

I awoke early to another wet Sheffield day. I had no new bites or rashes; if there were bed bugs in that dank hostel, they apparently did not find me. I dressed and packed my things, and then went down to the kitchen and made a cup of tea amongst scattered beer bottles and dirty dishes. After bumming a ride to the train station from a nice Englishman, the only hostel resident other than my Shanghai roommate who had not been at the brandy party the night before, I boarded the 11:27 a.m. back to London. As before, I spent much of the two-hour ride thinking about my overnight backpack on the train's luggage rack, nestled in with several others, and all of the places I had traveled on my adventure so far. The rash on my side, though quite possibly not from bed bug bites after all, itched once more.

Back in London, I killed time at my hostel while I waited for David Cain to contact me for our meeting. I sat in the lobby reading through my notes and taking some more, contemplating whether I could stomach another night sleeping there. As I sat, a man came in to ask for a bed. "I'm sorry, we're full," replied the sympathetic Scandinavian woman at the front desk. The man left. Soon another man came in asking for a bed, with the same result. During the hour I sat there, around a dozen people came in looking for a place to sleep that night; one threw up her hands, complaining that there wasn't a single open hostel bed in the entire city. When she left, I asked the woman at the desk if this had been happening all day. "Yes, since morning," she said. I was stuck.

A text came through from Cain with directions to his office, along with a cryptic message that read: *Hope you were not at the Trop medicine school.*

Why's that? I wrote back.

The place they confuse dust mites for bedbugs. I'll show you TV footage, he responded, referring to a BBC news clip that I watched on YouTube several months later. While it was true that the footage showed one of the school's scientists peering through a microscope at what appeared to be dust mites, I learned that the news network had inadvertently added the incorrect images after filming.

After the Tube ride to Cain's office, our dig through his bed bug museum, my ride back to my hostel, and a brief curry dinner, I resigned myself to the fact that I had one more night spooning my luggage. I was in a different dorm than the one I slept in prior to my short trip to Sheffield, but it may as well have been the same room. As before, there were five bunk beds filled with young people from all over the world, who came in at various times of the night, stumbling drunkenly over the suitcases on the floor. There were snorers and thrashers and at least one person with very stinky feet. I rose before the sun, made one last trip to the plastic shower capsules, gathered my belongings, and began the long trip back to Heathrow via the metro and a train from Paddington Station. In just a few hours, I would land in Prague.

NINE

THE ORIGIN OF A SPECIES

The Bed Bug's Beginning

"They will piss on us," Ondřej Balvin had warned me. At the time, sitting in a dim bar in Prague's Žižkov neighborhood, the idea of little bats peeing on me while I collected bed bugs from a roost seemed funny. Less than twenty-four hours later with 1,260 bats poised over my head, I wasn't so sure. They were greater mouse-eared bats, or *Myotis myotis*, and they made up the second largest colony of the species in the Czech Republic, lining the wooden rafters of an apartment house attic in the town of Dubá, roughly forty-five miles north of Prague. The bats' sharp squeaking chorus and earthy ammonia stench made it hard for me to follow Balvin's running commentary on our assignment. Every minute or so, a bat ventured from the group to swoop the length of the attic. I jumped with each pass. I knew the bats wouldn't hurt me; they were interested in beetles, not people—and, besides, they were used to the occasional scientist climbing the steep wooden steps to count them or to collect the bugs that lived with them. Still, knowing a fact doesn't settle the reflexes. I tried to imagine a cave with hundreds of times the number of bats, wondering how I might react to the Kenyan caves on Siva-Jothy's adventure. I could not.

I stood a safe distance from the pee zone and adjusted a headlamp lent to me by Zdeněk Vitáček, a naturalist at the nearby Česká Lipá Museum and an official bat counter. Balvin handed me a pair of flimsy tweezers, a plastic bottle with a folded piece of filter paper cradled inside, and a small flip-top tube filled with alcohol. I juggled these along with my notebook, pen, camera, and recorder, unsure whether to report or play scientific assistant first. I eyed the chattering bats clustered on the ceiling around a chimney, which

Balvin said was the most likely place to find the bugs. I stuffed the collecting materials into the pockets of my jeans, opened my notebook, and jotted notes. Unaware of my internal struggle with the bat colony, Balvin strode to the chimney and began pinching bugs with his tweezers, and Vitáček shone a strong flashlight into the shadows of the pitched ceiling to start his count.

I had found Balvin, a bespectacled PhD entomology student at Charles University in Prague, through the team at North Carolina State University—he had sent the American team bed bug samples for their global population genetics survey. After months of e-mails and planning, Balvin had agreed to be my guide for much of my twelve-day excursion in both the Czech Republic and Slovakia.

Balvin had collected bugs from bat roosts similar to the one in Dubá for seven years in preparation for his thesis. At first, he was looking for *Cimex pipistrelli*, the bat bug, and had been tagging along with bat experts like Vitáček to find them. While the greater mouse-eared bat is not a threatened species, it is considered an ecologically important one throughout Europe, its native continent, and is protected under Natura 2000, a European Union initiative to preserve key flora and fauna. Around twenty people monitor the greater mouse-eared bats by counting specific populations and making sure the animals' habitats remain untouched, even when those habitats exist in occupied apartment buildings like the one in Dubá.

But Balvin found the common bed bug, *Cimex lectularius*, living in most of the bat roosts, despite the fact that this species was usually associated with humans. Even in the apartment roosts where the bed bugs could have fed on the people who lived just a floor or two below, they still chose the bats. The discovery shifted Balvin's research focus from bat bugs to bed bugs. Eventually he collected bed bug samples from pest controllers across Europe and, by a stroke of luck, collaborated with the researchers at North Carolina State University, even though he never met any of them in person. Genetic analysis of Balvin's Czech bugs, in part, led to the hypothesis that the pesticide-resistant bed bug strains may have originated in Eastern Europe, which was part of the reason I had traveled to Prague to begin with.

Zdeněk Vitáček counting bats in an apartment attic in Dubá, Czech Republic. Credit: Brooke Borel.

After about fifteen minutes, the sound and the smell of the bats shifted from sensory overload to background noise. I slid my note-book into my back pocket and stepped closer to the colony. From this view, I saw that the bats had fuzzy bodies and large rounded ears, and their wrinkled faces and delicate wings were pink. I pulled out my camera and began to take photos, no longer flinching when a bat flew out of the roost. ("If you pick them up, they will bite," warned Balvin, as I contemplated out loud whether they had rabies or other viruses.) I put the camera back in my pocket and pulled out my collecting materials—the paper-filled bottle for live bugs and the alcohol for dead ones—and stepped over piles of slick bat guano to examine the chimney wall.

From a foot away, I could see the coarse sands of the cement speckled with the dark brown flecks of both bat and bug feces and the lighter brown spots of the bodies of bed bugs squeezed into tiny cracks and holes. I moved closer, sweeping my headlamp back and forth, trying to catch the bugs in their hiding spots. There! I took my forceps and gently pinched a bug from a crack and brought it in close to see which container to put it in. The bug's tiny legs

Ondřej Balvin collecting common bed bugs from the chimney in the Dubá attic. Credit: Brooke Borel.

kicked in the air, so I popped the lid off of the larger bottle, shook the bug against the filter paper, and sealed it. Back at the wall, I found another bug. Its legs were still. Into the alcohol it went. My unsteady hands dropped the next live one into the layer of bat shit on the floor below and, not wanting to waste it, I hunched down to look for its alarmed scramble as it worked its way back to the wall. There! I pinched it up and flipped the lid off the larger container so it could join its friend.

Finding the bugs became a game. The first step was to spot them by discerning a dark brown smudge of guano from a lighter-shaded insect. The next step was to guess whether the bug was dead or alive before I went in, because we needed more live bugs than dead ones and we didn't have much time. In the beginning, I couldn't recognize the signs of life, but after ten minutes I could see that the live bugs emanated a vitality that I couldn't pinpoint even when they were frozen in place. The live bugs were brighter and the pitch of their legs was more purposeful, ready to take off running. Once I found a live one and caught it between my tweezers, the next step

Myotis myotis, or greater mouse-eared bats, roosting in the Dubá attic. Credit: Brooke Borel.

in the game was to tap the collecting container against my palm to settle the captured bugs to the bottom, pop off the top, slip the new bug inside, and then snap the lid on. I settled into a rhythm: seek, pinch, tap, pop, drop, snap. And repeat. I shouted when I captured a bug and cursed when I lost one among the bat shit, which was soon ground into the deep treads of my hiking boots.

As my collection bottles filled up, I remembered a passage from a Charles Bukowski biography where the poet is thrown in Moyamensing, a low-security prison in Philadelphia, for dodging the World War II draft. The poet spent his first evening betting dimes with a fraudster named Courtney Taylor over which man could catch the most bed bugs. (Taylor allegedly cheated by tearing his bugs in half and stretching the pieces so they each counted double.) As I tried to keep pace with Balvin, the unwitting competition in my own bed bug game, I thought back to just two days earlier when, upon my arrival in Prague, I had used a flashlight to inspect the slats of my hostel-issued bunk bed. Later that day, after a couple of glasses of Gambrinus pilsner with the hostel's manager, I got up the nerve to ask about the faded flecks of bed bug feces I had found. He admitted the hostel had bugs in the past but insisted it hadn't been a recent problem.

A drop hit my head. Then another. It was bat pee.

• • •

Prague sits in the heart of Bohemia, the largest historical region in the Czech Republic. Bohemia's past extends back to the second century BCE. To the east is Moravia, another swath of old land, which Robert Usinger and other entomologists of the fifties and sixties thought had a natural divide separating the bed bug, *Cimex lectularius*, in the south from the bat bug, *Cimex pipistrelli*, in the north. Usinger had planned to visit his colleague, Dalibor Povolný, in Moravia in the late summer of 1968 to collect bed bugs along this apparent natural delineation to help prove this distribution of species.

The trip, if Usinger hadn't been forced to cancel it due to his failing health, would have been his second attempt at seeing Povolný in Czechoslovakia. The two men first met a decade earlier at a Royal Entomological Society meeting in London, where Povolný noticed a reprint of a paper he had written on bed bugs sticking out of Usinger's pocket. Usinger was taken with the research, which described the common bed bug living in the roosts of greater mouse-eared bats in the baroque Slavkov Castle and the tower of the pilgrim church in Křtiny, both in Moravia. The paper was the first documentation of this phenomenon, and Usinger took it as evidence that the bed bug had originally lived on bats before moving onto humans or our early relatives. In fact, he found the work so intriguing that he insisted on meeting his new friend in Moravia to see the bat-loving bed bugs for himself. He also wanted to research the supposed geographical line Povolný had discovered that separated the bug species.

In an interview decades later, Povolný claimed that Usinger had made the trip, under the condition that the Czech government's secret police could trail the scientists wherever they went. Letters between the two scientists, however, suggest otherwise—although the two plotted the trip for months, their plans fell apart because they couldn't secure the appropriate visas and other permissions from the communist country. Povolný's assertion, though likely

Dalibor Povolný (*far left*), Robert Usinger (*center*), and William China (*right*) at the Royal Entomological Society of London in 1958. Courtesy of Bancroft Library, University of California, Berkeley.

embellished for the interview, wasn't exactly outlandish. It was common for both Eastern and Western governments to track international scientists during the Cold War to make sure they were not truly spies. In 1963, for example, when Povolný took a working trip to Washington, DC, California, and elsewhere in the United States, the Federal Bureau of Investigation opened a file to follow the Czech entomologist's every move, even detailing when he checked into a room at a Berkeley YMCA "because he was very low on funds." The FBI also opened a file on Usinger in 1954, possibly due to his own international travels as he worked on his bed bug monograph.

Whether or not Usinger made that initial trip to Czechoslovakia, he and Povolný kept up correspondence over their decade-long friendship. In several letters, Usinger begged Povolný to send him bug samples from Eastern Europe to include in his research for *Monograph of Cimicidae*, and Povolný repeatedly captured what he thought were bat bugs from roosts in Moravian and Bohemian churches and attics, fed them blood from his own arm, and then sent them to Berkeley by international mail. Each time, Usinger would write back that the bugs were in fact the common bed bug, rather than the bat bug.

Starting in 1964, Povolný traveled to Afghanistan to help teach zoology to students at a new medical program at Nangarhar University in Jalalabad, as well as to work as the school's first parasitologist. During a four-month stay in 1965, Povolný managed to go on side trips to hunt for bed bugs, and he eventually found the common bed bug that typically feeds on humans living in a cave with *Myotis* bats that are a close relative to the greater mouse-eared bats in Europe. Povolný gathered some of the bugs and fed them on his blood to keep them alive until he broke out with blood blisters, after which he paid a local Afghani to feed them instead. He brought the bugs on his flight back to Czechoslovakia and then mailed them to Usinger in Berkeley.

After a tense moment when Plant Quarantine officials from the California Department of Food and Agriculture intercepted the bed bugs, the package arrived safely at Usinger's lab. Some were even still alive, and Usinger set to work on cross-breeding experiments to see if the bugs could mate with bed bugs that had been living on humans, rather than bats. If the bugs could reproduce and if their offspring were fertile, it was a hint that they may be the same species surviving on two different hosts rather than two distinct types of bugs. The experiments were successful, and Usinger wrote to Povolný: "The most fascinating part of this is that you have now filled the most important gap in the theory as to the origin of human bed bugs." Povolný's discovery in Afghanistan was perhaps the missing bed bug link: the first evidence confirming that the bed bug species that feeds on humans also lives with

bats in caves and not just in man-made structures. The fact that Povolný found the bugs in the Middle East, the supposed ancestral
origin of bed bugs, was a bonus.

In 1966 the two men jointly authored a scientific paper on the
findings entitled "The Discovery of a Possibly Aboriginal Population of the Bed Bug." The researchers noted that the common bed
bug's evolution might predate that of modern humans, and they
were able to tentatively show that the insect originally lived on
bats and moved to humans. But the only tools they had available
were comparing the shape of one bug to another—and, in particular, differences in the shape of the spermalege, the female's odd
reproductive organ, and the size of the bugs' heads—as well as examining the bugs' number of chromosomes, the DNA-containing
structures inside the nucleus of most cells. The findings hinted
that their hypothesis was correct, but they admitted that further
evidence might prove or disprove their work.

Had Usinger survived into his nineties, perhaps he would have
been disappointed to learn that the invisible Moravian line separating the common bed bugs from the bat bugs was proven wrong
in 2006 when Balvin discovered bed bugs and bat bugs on the
wrong side of the supposed border (although, had Usinger been
healthy enough to make the trip, he may have discovered this himself). More likely, he would have been happy to learn that the common bed bug still thrived in bat roosts like the one in the Dubá
apartment attic, and that a young scientist was still gathering the
bugs to further unravel their beginnings. Maybe, too, he would
have marveled at the population geneticists who were trying to reconstruct the DNA of those bed bugs and others scattered across
the globe to map the insects' relationships.

• • •

Back in Prague, curious to uncover the origin of the more recent
bed bug trajectory—the spread of the pesticide-resistant bugs—I
visited the National Institute of Public Health, a sprawling compound on a residential street on the city's east side. There, I met
with František Rettich, the director of the institute's National
Reference Laboratory for Rodent and Insect Control. I trailed him
through the department's long tiled corridor to the insectary,
which smelled like spoiled milk, a sourness that spilled out of the

room and into the hallway. I couldn't tell where the smell came from, but there were a lot of possibilities. Long metal shelves on one side of the room were lined with thick glass jars that held German cockroaches, American cockroaches, and the famous hissing Madagascar cockroaches like a nightmare candy display. The hissing roaches were kept not for research but "just for fun," said Rettich, as I peered through the glass at their impressive frames. Unlike the pet roaches in David Cain's office in London, these were in full view. Their thick and shiny articulated bodies, each longer than the palm of my hand, were frozen on the floor of the jar like mechanical toys in need of a windup.

Identical shelves on the opposite wall held lidless glass aquariums with ants so tiny I had to squint to be sure that they had legs, as well as netted boxes of various flies and mosquitoes. The open ends of the wispy netting were tied up like hair buns, which the researchers would unknot when they needed to reach the insects inside. Three ribbons of flypaper studded with carcasses curled beside an enclosure that was dense with black flies, which I was told are especially hard to contain even with netting.

Tucked in a corner, away from the glass displays, sat seven plastic cups sealed with netting and rubber bands. The cups held five populations of pesticide-resistant bed bugs, four from homes across the Czech Republic and one from an American hotel. Zdeňka Galková, a graduate student dressed all in white from her shirt to her socks and sandals, snapped a rubber band off one of the cups and showed me the young straw-colored bed bugs inside. Galková had been working with the bugs for just four months and hadn't yet had time to transfer them to permanent housing, a stack of seventeen glass jars on the shelf below. The number of jars seemed optimistic. Before she had enough bugs to fill so much glass, she had to build their numbers—a tricky endeavor in a lab with no animals to provide the blood. The plan, she said, was to develop an artificial feeder. I asked what the bugs ate in the meantime, and she rolled up her pant leg to reveal a faint pink mark from their last meal four weeks before, roughly the diameter of the mouth of one of the cups.

Once Galková had enough bugs, she planned to test pesticides on them as part of the lab's main line of research, which is

The insectary at the National Institute of Public Health's National Reference Laboratory for Rodent and Insect Control in Prague. Credit: Brooke Borel.

to evaluate product efficacy for registration and labeling. Under the European Union's rules, pesticides go through a registration process very similar to the Environmental Protection Agency's in the United States, with a notable difference. Citronella and its "all-natural" cousins do not enjoy a loophole with the EU, and companies that would like to register these must pay a fee of roughly $500 and have them tested for efficacy just like any other product. Rettich explained that he advises these companies not to ask his team for product testing. He knows from experience, he said, that the oils and essences don't do much more than make the bed bugs smell good. "If you insist, I'll do it, and you will pay a lot," he said. "Please, save five hundred dollars because I will likely say no in the end."

Down the hall from the insectary in Rettich's office, a high-ceilinged room with walls painted an institutional cream, I sat with the sixty-seven-year-old entomoparasitologist to learn about bed bugs in Prague. ("I should be retired four years ago," he told me, "but, still, I work.") Like Galková, he was dressed in white, a uniform I learned is an institute-wide safety requirement to make it easy to spot spilled chemicals or blood.

As a vector control researcher, Rettich mainly works with blood-sucking arthropods such as mosquitoes and ticks. But as the bed bug problem in Prague surged over the past decade or so, he lamented, his responsibilities extended to this pest as well. As we settled into our conversation, he told me he first "met the problem of bed bugs in the mid-eighties," when a friend asked for help getting rid of beetles in his apartment. Although Rettich had never seen bed bugs in person, he knew immediately that his friend was not suffering from beetles. Still, the bug was exotic back then—when head lice and cockroaches were more common Czech pests—and Rettich was able to clear up the susceptible bugs easily. "I sprayed my friend's flat with the pyrethrin. And it was perfect."

The narrative that followed fit other Czech stories I heard during my trip. The bugs were widespread up through World War II, after which they were controlled, in part, by dragging furniture out of homes and burning it. After the war, there was a respite that was partially aided by DDT and other insecticides. Then, in 1968, the Soviets invaded Czechoslovakia to quash an attempt at reform and to whip the country back in line with the communist bloc. According to many of the Czechs that I interviewed, although Rettich did not agree, the Soviets brought more bed bugs, which terrorized the country until it was able to control the insects once again with DDT (the pesticide was banned by 1974). During the rest of the communist era, the bugs were rare, in part because travel outside of the country was strictly limited and the bugs were susceptible to pyrethroids. Then the bugs bounced back around 2006. The reasons the Czechs gave me were familiar: tourism, the purchase of used furniture, and insecticide resistance.

Prague seems as good a city as any in the region for bed bug ground zero. It is one of Eastern Europe's largest cities, with a bustling tourism economy that draws travelers from all over the world. But the timing didn't synch with the theory that the resistant bed bugs in the United States had come from Eastern Europe. If large numbers of the bugs started showing up in 2006, it seemed unlikely that they were responsible for seeding the resistant bugs that landed in America in the nineties.

Even if the bugs hadn't spread from Eastern Europe to Western Europe and into the United States, where had the Czech bugs come from? Rettich suggested that the resistant strains might have originated somewhere in Africa. Within Prague, he added, although he didn't necessarily agree, people blamed Chinese and Vietnamese immigrants for spreading the pests within the massive Soviet-era apartment complexes that dot the city. To Czechs, he said, these foreigners have "strange habits," including a higher tolerance for pest insects. Czechs also blamed the Roma, more commonly and pejoratively called gypsies, a growing and impoverished ethnic minority throughout Eastern Europe. Most live on the fringes of society, at the edges of cities or in shantytowns in the countryside, although some apartment buildings within Prague are mainly Roma, which means a possible physical connection on public transportation and in other shared spaces.

Rettich led me back to the tall gates of the institute. As I began my walk back to my hostel, I wondered whether hunting the resistant bed bugs' origin was as futile as my internal debate over whether I had been bitten by bed bugs in London. Each person I spoke with told a story with a similar arc, but with different characters: the bed bugs fell at some point in the mid-1900s, then rose half a century later, and the people or regions who were to blame were usually outside the demographic of the person doing the theorizing. It wasn't surprising, considering the human tendency to label *us* versus *them*, or *familiar* versus *foreign*, especially when it comes to disease or infestations. This held true even for a bug with a history so intimately intertwined with our species as a whole. Bed bugs evolved to survive on any human blood, and yet they are always thought of as belonging to someone else.

• • •

The pest controllers I met northwest of Prague blamed the bed bugs on the junkies. In a towering apartment complex in Litvínov, a small town near the German border, I trailed three women who were speaking in rapid Czech to Ondřej Balvin and struggled to get him to translate for me. One was the building's manager, a short wide-eyed woman who carried a thick stack of papers in a blue plastic basket. The other two came from the building's state-hired pest management company. As we strode down a long hallway, Balvin told me that the manager had said the junkies used to live in this very wing of the complex, where they spent most of their time cooking crystal meth and shooting up. They had since been evicted. In recent months, the manager kicked out a particularly difficult tenant from a cramped studio at the end of the hall that overlooked lush green forests interrupted by a sprawling oil refinery. The room had been so bad that the cleaning crew had swept out towering piles of dead bed bugs, the manager said, throwing her hands apart like a bragging fisherman. Down the hall in another former junkie's studio, which had gold graffiti that read "Life Is Party" and "Back to Underground" in English, the walls were speckled with bed bug feces.

The manager unlocked the door of an apartment and stepped aside to let in the exterminators. I watched, mesmerized, as one of the women, a seventeen-year veteran named Hana Smíšková, treated the room. She was unlike any pest controller I had seen: short funky hair, heavy eye makeup with black fake eyelashes, disco ball earrings, a red company T-shirt paired with faded black jeans, red nails, and thick silver jewelry. A face mask hung uselessly around her neck, although she had protective booties pulled over her shoes and latex gloves on her hands. With a wide grin, she hoisted a canister of K-Othrine—a pyrethroid produced by Bayer CropScience, home of robotic chemical libraries and testing rooms—onto her back like a knapsack and doused everything in sight with a long metal wand until the room dripped from the ceiling down to the worn wooden floor. I asked if this was how all the treatments went, and as Balvin translated my question, Smíšková looked over her shoulder to answer as she blindly sprayed the wall in front of her. Yes, this was how they did it. (Months later I would ask an American bed bug expert how common it was to spray a room until it dripped. "More common than it should be," he replied, "but you simply cannot pussyfoot around with bed bugs.")

Smíšková's teammate for the day was Lenka Kučerová, who would soon inherit the company the two women worked for, a family-owned business run by Kučerová's mother, who had kept it going after her husband had died years before. Smíšková and Kučerová doused eleven apartments in about an hour and a half, usually spending less than ten minutes in each. My own bed bug treatments in New York, as well as others I'd heard about or observed across the United States, took about an hour. Unlike the two junkie apartments, most of the units were occupied. The pest controllers sprayed their pesticide thoroughly and quickly, soaking beds, blankets, couches, pillows, walls, and floors. When they ran out, they mixed some more by sprinkling the white powder into their canisters and topping them off with tap water sloshed from a red plastic bucket.

In one apartment that smelled of cigarettes, they found an old woman's pillowcase crawling with bed bugs. Balvin rushed in with a plastic collecting tube and tweezers, capturing around a dozen of the plump bugs before Smíšková and Kučerová could spray them. The exterminators spoke to the old woman in Czech as they stuffed the pillow into a flimsy trash bag that was too small to tie off. "They're telling her they need to throw it away immediately," Balvin translated for me. The woman nodded and then showed us the bites on her arms as she described her pain in a husky smoker's voice.

Across the hall, a pinched-faced woman opened the door, her

Hana Smíšková treating an apartment for bed bugs in Litvínov, Czech Republic. Credit: Brooke Borel.

young son peering around her legs. When she saw my notebook and pen, the woman cursed the building manager in a shrill voice. After negotiations that I did not understand, she moved aside to let us all in and then stood with her arms folded as the exterminators showed me bed bug flecks scattered along the edges of the bed that stood in the center of the apartment's single room. Plastic toys and clothing were stacked several feet high around the studio's perimeter; I grew dizzy trying to count the number of places a bed bug could hide. The woman pointed to a wooden dresser, which she said was the source of the bed bugs. She had bought the dresser from a neighbor.

When we left, Kučerová told us that another company had treated the woman's apartment six times already with a pyrethroid mist, but that her company's process would work better. It was, after all, working in the other apartments in the complex, according to the highlighted map that the building manager carried in her plastic basket. Out of around 350 units, there were 31 with bed bugs, and the manager agreed that the bugs were getting much better since Kučerová's company had been working there. As I took notes, I wondered how long that would remain true with a pyrethroid that was showing wide resistance worldwide.

The company, one of just two like it in the nearby town of Most, had its first bed bug case in 2006. The bugs make up 90 percent of the women's exterminating jobs, and they do around 2,000 treatments each year including repeat jobs that didn't work the first time. They told me their success rate is good, and that the pyrethroid-dousing approach is working for them. When I asked if they collaborate with any local entomologists, geneticists, or other scientists, however, I learned that there is no opportunity for such cross-pollination. Instead, the only education Kučerová and her team receive comes from industry meetings sponsored by the chemical companies, which provide instructions and application techniques for new products.

The approach is not unusual: as in the United States, pyrethroids are the main option for indoor use in Czech Republic, although the latter is still able to use one carbamate. Back in Prague the next day, I would watch an independent self-proclaimed bed bug specialist methodically spray a pyrethroid on the mattresses, blankets, and frames of three beds in a large railroad apartment occupied by a Roma man and his two adult children. Again, the infestation was blamed on a junkie—in this case, the man's son, who frequented local hostels and was addicted to meth. Whether his drug use or hostel visits were relevant or not, his bed was the only one to turn up any bed bugs, which Balvin collected into a plastic container.

And later that week at a brief interview at a shopping mall coffee shop, another Prague pest controller, a tall grizzled man, listed out three pesticides he uses for bed bugs. The first two were pyrethroids. The third was his secret weapon, which he told us he uses for around 10 percent of his seventy or so bed bug cases each year. It was a carbamate, like propoxur, the pesticide that Ohio is fighting for permission to use in the United States. That one worked. For now.

• • •

Curious about the bed bug situation farther east, Balvin and I hopped a rickety train to Košice in eastern Slovakia—the other half of the former Czechoslovakia, which had separated from the Czech Republic in the early nineties in what historians call the Velvet Divorce. I also wanted to learn whether Roma settlements really had bed bug problems and, if so, where *they* thought the bugs came from. There are more Roma in Slovakia than in the Czech Republic, and Balvin had connections at two settlements on the outskirts of Košice. While the local government officials who oversaw these communities thought our request to visit was strange, they said we could come.

I spent the long overnight ride on a hard and narrow bunk with pillows and blankets borrowed from the conductor, who insisted that the train never had trouble with bed bugs. After a restless sleep interrupted every hour or so by garbled announcements at tiny stations along the way, we arrived in Košice in the early morning, walked to a nearby bus terminal, and, after some confusion over the timetables, boarded a bus headed for the small village of Nováčany. The village was old, dating at least to the 1300s, and its streets and gardens were quiet and neat. We walked to a yellow building where Balvin had been told we would find the mayor, who had agreed to take us to a nearby Roma settlement.

The mayor and her assistant were waiting for us, two middle-aged women in fashionable slacks and blouses with coiffed hair. They smiled warmly but neither of them spoke English and I knew no Slovakian, so I was again dependent on Balvin's translations. (Czech and Slovakian are similar enough that speakers can understand one another.) After brief pleasantries, which for my part involved nodding and smiling, the women tossed light jackets around their shoulders and then led us out of the office, down the road, and toward the Roma camp.

As we approached the camp, I saw that it was a cramped cluster of shacks built from scrap wood, stone, and bricks with sloped roofs of corrugated metal. Around two hundred people lived there. Satellite dishes on tall wooden poles stuck up from some of the buildings like gray lollipops. We entered via a steep dirt path. Sharp smoke from wood-burning stoves tickled my nose, but it didn't cover the underlying decay of open sewage. As the mayor began knocking on doors and people peeked out of open windows at our intrusion, I wondered why we had come. The idea of asking these people, who are among the poorest and most disenfranchised demographic in Europe, if they had bed bugs no longer seemed like an exercise in curiosity or journalism but one in insensitivity.

But it was too late and so I followed the mayor from door to door, scribbling notes as Balvin translated pieces of the conversations. We learned that while a handful of people in the village had gotten bed bugs in recent years from furniture they had bought second-hand in Košice, there hadn't been widespread bed bug problems in the camp for thirty or forty years. It would be unlikely for bed bugs to come in or out by hitchhiking on luggage. While a few people traveled—including a young man who told me in perfect English that he went to school in the UK and was training to be a soccer coach—most stayed close to home. It would also be unlikely to bring bed bugs to or from work or social outings. The Roma here, the mayor told us, survived almost solely on government welfare, partly because white Slovakians would not hire them, although she said she sometimes tried to find them work in Nováčany. Slovakian schools, while sometimes open to both white and Roma children, are also notorious for segregation in the classroom and lunchroom as well as on the playground. Without direct interaction, bed bugs have no way to pass from one community to the next.

The next day we visited a sprawling Roma housing project called Luník 9, the largest in the country with between 7,000 and 14,000 people, depending on the season, crammed into apartment blocks built to hold only 2,500. Even before we arrived, people were shocked to hear we planned to visit the projects, warning us that

Roma village outside of Nováčany, Slovakia. Credit: Ondřej Balvin.

we'd be robbed or beaten. The owner of the small hostel where we stayed in Košice, initially friendly and talkative whenever I passed him on the narrow stairs, stopped speaking to me after I told him I planned to spend time with the Roma. But what I found at Luník 9 wasn't what people had warned me about. It was true that the projects were economically depressed, with blown-out windows and drifts of garbage more than six feet tall, biannual hepatitis outbreaks that happened like clockwork, and rationed water that was turned on just twice a day. It was also true that most of the inhabitants were squatting in the projects illegally, and that some of the high-rises were near collapse. Still, there were also on-site social workers and a doctor, as well as municipal workers and police. There was a nursery school and a few convenience shops. And while the people who lived there eyed us when we stepped off the bus and walked to the mayor's office, their stares were not hostile but inquisitive.

The mayor wasn't there to meet us like he had promised, and the doctor made his nurse shoo us away. The social workers were nice but said they weren't really supposed to talk to reporters. They suggested to Balvin that we could try a municipal worker, a blond

Luník 9 outside of Košice, Slovakia. Credit: Ondřej Balvin.

woman in a bright green vest who was across the wide courtyard supervising men as they shoveled raw sewage into a dumpster. The social workers also told us that we were welcome to walk the grounds and talk to anyone who was willing.

Although it was a weekday, the Luník projects were packed. Everywhere I looked, people strolled aimlessly through the yard or hung out their windows or kicked soccer balls or smoked cigarettes. When we spoke with the municipal worker, her voice rose as she explained, in Slovakian, the poor state of the buildings and the clouds of fleas. She said the government did what they could—in addition to the annual sanitation cleaning that she was supervising, the state also provided insect control once a year and rat control twice a year. But it was not enough. As far as employment, the situation was like Nováčany, but worse, with nearly all of the residents surviving on government subsidies with little opportunity to work. After forty-five minutes, she ran out of complaints and sent us to walk around.

We spoke to a family sitting on a curb, and the toothless father laughed when Balvin asked about bed bugs. *That* was why we were there? No, he said, they never had bed bugs, and then he pointed across the courtyard to another building. *Those* people did. We walked more, stopping by a group of girls bouncing children on their knees, who said they had no idea what a bed bug was. After stopping to talk with several more groups that sat in the yard, as well as laughing with kids who ran up, shouted questions, and then ran away, a group of men sauntered over. They were led by a pot-bellied man in an Adidas shirt, who I later found out was an unofficial Luník 9 representative. He drilled us with questions. Who were we? Why had we come? Balvin explained, and the man nodded and said he'd tell us the real story of the bed bugs. The Roma didn't have bed bugs naturally, he said, but instead picked them up when buying used furniture from thrift shops or bazaars. The furniture came from the white Slovakians, he said, never Roma. It was the white Europeans who were responsible for spreading bed bugs.

Later that evening at a picnic bench outside of a beer hall, I tried to find out where Maroš Bačo, a Slovakian exterminator, thought the wave of bed bugs over the past decade had come from. Before I could, he turned to me and asked:

"Do you think the bed bug is an American bug? Do you think everything started in USA and spread to Europe?"

"Is that what you think?" I asked.

"I don't know exactly."

I explained the different hypotheses by the population geneticists, entomologists, and pest controllers in the United States, adding that some people thought that the pesticide-resistant bed bugs might have even originated in Eastern Europe. Bačo then told me that he had been in the pest control business in Slovakia for twenty-one years but had only seen his first bed bug around 1999 or 2000. That year, he had to treat just two or three apartments in Bratislava, the capital city. He used a pyrethroid, which he said he learned about by reading online forums from pest controllers in the United States, and "it was fantastic." The next year, he had between twenty to fifty bed bug cases, and the chemicals still worked. The next year, there were more cases and, still, the chemicals worked. Now, he said, there were hundreds of cases a year. And the pyrethroids had stopped working.

The bed bugs were increasingly infiltrating Bratislava, which had a more established flow of tourism than Košice. Even several new and expensive hotels were filled with bed bugs.

"If they have tourists from America, they have more bed bugs," he said, grinning.

"See, you do think they come from America," I said, laughing.

"So, we had a few treatments in hotels. Maybe sixty, seventy percent, it was after an American stayed there."

• • •

"Shit, one is missing," said Balvin. I stopped taking notes and stared at his arm. We were back in Prague on my last day in Eastern Europe, and we were sitting in his office at Charles University with the bed bug catch from the past week and a half spread out on a small table in various plastic tubs. There were the dozen or so from the old woman's pillowcase in Litvínov, another ten from the Roma family's apartment in Prague, and more than a hundred from the bat roost in Dubá. The bugs from the apartments were fat and happy, having recently fed from their sleeping hosts. The bodies of the bugs from the bat roost, however, were flat. They weren't quite starving, but they needed to eat soon. Balvin had considered asking a classmate to borrow one of her bats for the feeding, but he had lost his nerve. The backup plan was human blood, which I politely declined to provide, citing my previous allergic reaction, infection, and subsequent emergency room visit. And so Balvin learned to feed bed bugs on himself for the first time in seven years of studying them. "No. Shit, it's here," he said. *Phew*, I thought.

We'd already had another near-escape. Balvin's professor, who shared the office, had given him a plain glass cup meant to hold a tea candle, which the lab sometimes uses in place of petri dishes. Balvin tipped a few bed bugs into the glass, pressed his forearm on top, then flipped it over so that it was resting upside down. On the first try, a bug snuck through a small gap between the glass and his skin. After ushering the loose bug into one of the plastic containers on the table, his professor secured the glass in place with a strip of yellow tape.

"Can you feel the bites?" I asked.

"No, just their feet," he said.

I squirmed in my seat. The constant movement in the containers laid out in front of me, which I could see out of the corner of my eye even as I bent over my notebook to write, made me more nervous than I expected; the bugs endlessly crawled on the crumbled pieces of tissue paper that served as their temporary refuges, a constant background motion. At the time it didn't occur to me that the bat-feeding bed bugs might harbor a pathogen capable of infecting a human—it was just their presence that was unsettling. After all, decades of research said that bed bugs weren't able to spread a variety of bacteria, viruses, and parasites, and medical entomologists had been telling me for several years that the bugs weren't dangerous. Four months after I watched Balvin feed the bed bugs from the attic bat roost on his arm, however, I would get a sinking feeling as I read the new paper by researchers at Virginia Tech suggesting that bed bug experts might not be looking in the right place when it came to bed bugs and disease. Their hypothesis, of course, was that the most likely place for a bed bug to pick up a virus was a large population of non-human animals. A flock of birds was a possibility. So was a roost of bats.

After reading the paper, I searched for an expert who could tell me whether or not the Dubá bug-hunting adventure could have launched the next worldwide mystery-virus outbreak. I ended up on the phone with virologist and wildlife disease expert Simon Anthony from the Center for Infection and Immunity, who took me through the gymnastics that a hypothetical virus would have to maneuver to cause a catastrophic outbreak. First, Anthony said, there were the biological and physiological barriers. Before a virus could spread from a bat to a human through a bed bug bite, the bug would have to be capable of pulling up the virus in a blood meal and ingesting it. Once inside the bug's digestive tract, the virus would have to be able to bind to the gut cell walls, escape those cells, travel to the bug's salivary gland, and then exit the mouth during the next blood meal. While there are a lot of blood-borne viruses that can spread this way, it is still rare for the microbes to have the right structure and ability to infiltrate any given organism.

Even if a bat virus could transmit to a bed bug and end up in its spit, other obstacles await in the very first human body it encounters, or patient zero, in which the virus may not be able to establish infection. And even if it does, the virus could still find a dead end.

In order to transmit to a wider range of people and thus succeed at cornering a new market, the virus would need to be able to transmit from large numbers of bed bugs to a densely populated group of people. Considering most individual bed bugs live in one place and feed on just one or a few people, as well as the fact that most people don't dwell in highly dense homes with many others, even a viable virus is not likely to spread very far. "So you're really talking about the perfect storm here," Anthony said to me. The bed bugs in the attic weren't it.

Australian entomologists took the idea even further in a 2012 publication. Their back-of-the-envelope calculations suggested that the bed bug resurgence in the United States has included at least 220.5 million bugs since 2000, with no obvious outbreak of disease. The scientists concluded: "It is not possible to prove that bed bugs cannot transmit any infectious agent; however . . . the indications are that the risk of contracting an infectious disease through the bite of a bed bug is almost nonexistent."

Back at Charles University, Balvin finished feeding about fifteen of the bed bugs from the bat roost on his arm, sorted them into two plastic test tubes, and then handed them to me for labeling. The lids of the tubes were built in the lab, with a hole cut out of the top and replaced with a piece of thick netting so that the bugs could have some air. I picked up one of the tubes and rolled it back and forth between my forefingers and thumbs, watching the bugs as they huddled in the folds of a twist of tissue paper inside.

These bed bugs wouldn't necessarily help uncover the origin of the pesticide resistance that had infiltrated bedrooms worldwide. But, like those Povolný had found living on bats in Afghanistan in the sixties, they may have been another link in the human bed bug's ancestral past. Balvin's preliminary data from earlier catches suggested that the bugs had divergent genetic profiles compared to samples of the same species that fed only on humans. The two strains weren't genetically different enough to be considered separate species, although early tests did not conclude whether they could successfully interbreed. It also wasn't clear whether the human-eating bugs originated from the bat-eating bugs, or vice versa. Generations may have been living on the bats, and then turned focus to humans sleeping in the apartments below. Or, perhaps, some of the common bed bugs living in the Dubá apartments made their way to the attic, drawn by hundreds of warm bodies, and stayed.

"Are you labeling?" Balvin asked. I picked up a pencil and marked the collection location, the number of males and females, and a custom number code onto sticky paper labels while Balvin finished his feeding and packed a few more tubes. In the end, there were fifty-eight bugs inside of six tubes: one from the pillowcase, one from the Roma apartment in Prague, and four from the bat roost. Balvin opened a cabinet to retrieve a small padded envelope, dropped the six tubes inside, and then poked a half dozen holes through its sides. The bugs were about to start their own voyage to the United States, where they would join the growing collection at North Carolina State University. There, perhaps, their genetic material would help scientists piece together the rest of the bed bug's story. Their genes may also reveal that the bugs were the start of a new bed bug strain, splitting from humans to specialize on bats. If so, it would be a poetic return to the original host that first passed the bugs to our ancestors during the Pleistocene.

• • •

Nearly six months after my trip to Prague, I called Warren Booth at the University of Tulsa, where he had moved on to an assistant professor position after his postdoctoral research in North Carolina. I was trying to get my mind around the bed-bugs-on-bats versus bed-bugs-on-humans story. Did the data from the Dubá bed bugs help clarify where the bugs came from? Could he see whether the pests were moving from humans to bats or from bats to humans?

"I think you can, with these data," he said, and described a three-dimensional map of genetic information from both the bed bugs from bat and human hosts, which he hung above his desk as he tried to understand its patterns. The work was preliminary and not formally published, but it came from more than just the bed bugs from the bats in the Dubá attic. There were data from 115 samples from more than two dozen similar roosts across Europe, as well as 99 bed bugs that came from 48 homes across a comparable region, where they had fed off of humans. The map showed the bugs' microsatellites, which are repeated messages within the DNA and show how a single bed bug is related to other bed bugs in a sample. The genetic information on the map separated into two distinct clusters that perfectly matched the bed bugs from the bats and the bed bugs from the humans, but with a significant difference: the data points for the human bugs were densely packed, while the bat-fed bugs were more dispersed.

"We found that the bat bugs have a much higher genetic diversity across the populations compared to human populations, and much lower levels of relatedness between the bugs from the bat colonies and from the humans," Booth continued. In other words, the bed bugs living with the bats did not share much genetic information with the bed bugs in human dwellings, which meant they were not interacting and breeding. This is why the data clumped into two distinct clusters. These differences stretched far back in time, according to the data, possibly around 250,000 years ago. And the higher diversity among the bugs living with the bats, which was represented by the more diffuse data points on the map, suggested there had been a genetic bottleneck when the bed bug split off to live with people. The pioneers that first moved to the new host had to provide all of the genetic material for subsequent generations, which ultimately decreased the variety.

"So what we're seeing is that the process of the bed bugs splitting off onto humans is actually still happening?" I asked.

"Yes, and it's a very ancient process," Booth said.

If the data and analysis can be confirmed, this suggests that the bed bugs are not moving from humans to bats and starting a new species after all. Instead, they are shifting in the other direction. The findings continue the work of Usinger and Povolný from nearly fifty years earlier, providing the genetic support that the two men couldn't have come up with at the time because the required scientific tools had not been invented yet. Booth and Balvin's team showed that the common bed bug may indeed predate humans. If the genetic data were correct and the insect split from bats to a human-like host around 250,000 years ago, it did so 50,000 years earlier than the first modern humans emerged in Africa. Around that time, the bugs may have lived in caves with both bats and an earlier human ancestor, and only later were the insects introduced to Homo sapiens. Then the members of the bed bug species were separated as some were whisked away to camps, villages, and cities, and the rest were left behind in the caves.

Over time, each set of bed bugs evolved unique mutations, which eventually led to behavioral, ecological, and physiological differences. Without migration between the caves and the modern human dwellings, the bugs' gene pools were increasingly isolated. Even today, though the bed bugs are found in bat roosts located in buildings that are also occupied by people, like the attic in Dubá, the bat-feeding bugs do not venture down to the apartments below or interact with any bed bugs that may be living there. Instead, the bat-feeding bugs likely moved in with their host when the bats discovered that an attic—or a church or a castle—was as good of a place to call home as a cave.

Despite this separation, the distinct strains of the common bed bug haven't been isolated long enough to separate into two different species. But we may be observing its evolutionary split onto humans in real time—the bugs living in our homes might be in the early stages of forking off from their bat-feeding brethren like a branch sprouting from a trunk, forging a new path in the tree of life. The version we struggle with today is just getting started on an evolutionary scale, regardless of the fact that we likely first met it during the Pleistocene; that it slept in the beds of tomb builders in ancient Egypt; that it radiated from the Mediterranean region to the rest of the world through all of our conquests and travels; that it faced catastrophic assault by DDT and then rose again; that it permeated modern homes and research labs; that it spawned an entire niche economy; that it has pushed people to suicide; and that it has worked its way into our music, literature, and art. Despite all of this, we are still watching the bed bug truly become *ours*.

More elusive than the bed bug's ancestral origin is the answer to my original research question: Where did the bed bugs that infiltrated the United States a decade ago come from? Including the bugs that showed up in my New York apartments? For this question, Booth's results were discouraging. His team analyzed the knockdown resistance, or the genetic mutations linked to pyrethroid resistance, from the bed bugs pulled from European homes and found that they are different from the same type of mutation in American bed bugs. This suggests that neither region is a major source for the other's bed bugs. Maroš Bačo's Slovakian infestations, then, may not have been seeded by American tourists, and Eastern Europe may not have been ground zero for the worldwide resurgence.

Still, at this point there is not enough genetic evidence to say for certain, and Booth maintains that Eastern Europe could still be the origin. Other researchers, including medical entomologists from Westmead Hospital in Australia, prefer the hypothesis that the resistant bugs came from Africa. To know for sure, the scientists must map the genetic flow of bed bugs from all over the world, which will require many more bed bugs than those that they have collected and analyzed from a few museums and apartments in limited global regions. Or they also could hit on the origin or origins quickly in an immense stroke of luck by collecting the right bugs from the exact source of the resistant strains. The genetic material from these bugs would need to be added to the database, analyzed, and double-checked. And as with the fifty-year gap between Usinger and Povolný's work on the bed bug's ancestral origin, perhaps the research on the resurgent bed bug's origin will benefit both from reflection and the boost of as-yet-unknown genetic technologies.

I ended my conversation with Booth and hung up the phone, feeling both satisfied that Usinger's work had continued in some way and also a little disappointed that I had not been able to fully answer the questions that had sent me on my quest to begin with. Then again, dozens of scientists hadn't been able to do it yet, either. Science is a long and dynamic process full of twists and turns and surprises, and some of the most interesting questions take many lifetimes to answer.

Whether we will ever find the origin of the resistant bed bug remains uncertain. Scientific and logistic challenges aside, another major obstacle is funding. The bed bug doesn't get much attention from the major agencies that support scientific research, particularly the institutions that would fund work on an insect or relevant control methods, including the National Science Foundation, the US Department of Agriculture, the National Institutes of Health, the Environmental Protection Agency, and the Centers for Disease Control and Prevention. This lack of funding is partly because the bed bug is not a known vector for disease and partly because these agencies aren't yet convinced that the bug is an organism interesting enough to study on its own merit.

This might change as more hypotheses on bed bugs and arboviruses emerge, to explore whether or not the insects are more of a threat to human health than is currently known. Or as the bed bug's ongoing journey toward a new species becomes clearer, institutions may deem them a more valuable research subject. Although speciation is likely happening all the time somewhere near us, it's unusual to catch it in the act; the bed bug may provide an interesting model to watch the process unfold. And as with all research, there is also the simplest drive: our curiosity. Isn't the desire to know how something works reason enough for scientific inquiry?

Regardless of the origins of the bed bug—both the deep ancestral dawn and the more recent comeback—one truth my journey clarified for me is that this pest belongs to all of us. We collectively helped shape the bed bug's very existence, both at the beginning when we carried it out of caves to our homes and then when we brought it across the world. The bed bug evolved to survive on us, adapting to the characteristics unique to our own species and exploiting the fact that we are social creatures by moving between our homes as we travel and interact. In modern times, we further sculpted the bed bug's very genes by forcing synthetic pesticides on it, helping breed the resistant strains that haunt us today. We hate the bed bug, and yet we inadvertently created it.

This pattern isn't unique to the bed bug—it's true of most modern plagues. We are at constant odds with insects and other pests that go after our bodies and crops, diseases that threaten our health, and parasites that sap our energy by feeding off of it. Yet these infestations and epidemics couldn't exist without us. They have evolved to use our agriculture, our homes, or our bodies in order to survive. We make and maintain the source of these plagues' ecological niches and at the same time try to kill them off. Mainly, we fight with an arsenal of increasingly sophisticated scientific and technological tools. Sometimes these tactics are very clever, like exploiting a bed bug's own biology to kill it; but we also try approaches that we didn't think through, such as blanketing the world in DDT. Both bed bugs and we humans are just trying to carve out our existence, to make a comfortable and livable place for ourselves. And it is a dance that will likely continue until the end of our time.

EPILOGUE
Horror, Curiosity, and Joy

Two weeks after I submitted this book to my publisher, I went to a conference in Chicago. It was the annual meeting of the American Association for the Advancement of Science, the largest conference for the largest professional general science society. In addition to thousands of scientists and hundreds of talks, roughly four hundred journalists converged on the meeting both to report on new research findings and to network. As I finalized my travel plans, I felt relieved that I would have a break from bed bugs for at least a couple of months as my editor and reviewers sorted out their revisions.

A snowstorm in New York delayed my flight and I arrived late to my hotel room, which I was sharing with a friend and fellow journalist. As I unpacked my bag and got organized for the meeting, we chatted about our plans for the following day and caught one another up on projects we were working on.

"I turned in my book," I said.

"Bed bugs! I'm terrified of getting bed bugs in a hotel room," my friend said. "I've never seen one before."

I laughed. "Well, I doubt you'll see any here. What are the chances?"

Exhausted from traveling, I soon crawled into bed and fell into a deep sleep. I awoke before my alarm went off. The sun hadn't even had a chance to creep in through the stubborn crack between the sliding curtains. Still foggy, I stretched and rubbed my hands together. *Ouch.* My index finger on my right hand hurt. It itched, too. It was familiar, a pain I'd experienced before. My eyes opened wide. *No.*

I shook my head as if I could will the feeling away. The movement made my neck hurt, and it was the same deep pain I felt in my finger. I raised my hand to the back of my neck to find it hot and swollen. My fingertips could detect small welts running down my skin, which I could read like Braille. *You bastards.*

I looked at the clock. It was 5:30 a.m. My roommate was a still bump under the covers in the room's other bed. I threw off my blankets and tiptoed to the bathroom, quietly shutting the door. I closed my eyes and turned on the lights. After a few seconds, I opened them and stared at myself in the mirror. My neck was noticeably puffy on the right side, and I had red splotches on both arms. I leaned toward the mirror and twisted my head around and inspected the welts on my neck. The bites. Let's just call them what they were.

I didn't have to be at the conference for another four hours, but there was no way to sleep any longer. I went back into the room and called my roommate's name.

"Sarah? Sarah?"

She stirred.

"Sarah, I am so sorry to wake you up so early, and like this," I said. "We have—I'm pretty sure we have bed bugs."

She sat up and stared. "What? What? Are you sure? How? What?"

In the end, I was right. The hotel management sent two men upstairs to our room. They tore apart my bed, took an official statement, and photographed my neck, arms, and hands. We found one bed bug during their search, a plump female full of what was most likely my blood. One of the hotel employees dropped it into a paper cup.

"You won't believe this," I said as I looked at the bug crawling around the bottom of the cup, "but I literally just wrote a book about these things."

"No kidding," the man said.

The other man got us a new room, took our laundry with a promise it'd be done free of charge, and gave us tickets for the breakfast buffet. Then, later that day, the hotel's head of security called me. There had been more bed bugs, he said, in the skirt of my bed. He told me the bed had been thrown out and promised me that the hotel would pay any medical bills and do anything else they could to make sure the rest of my stay was comfortable.

When I started this book, it was partly out of my feeling of horror and disgust toward the bed bug and its infiltration of my home. But I also started the project out of a deep curiosity. What was this insect and where had it come from? How was it that I had lived for nearly two and a half decades before I learned that it was a tangible, real animal? And if I knew the bed bug better—as best as I possibly could—would it help me control any lingering fears?

My experience in Chicago helped answer at least one of these questions. While I did not enjoy waking up in the early morning to discover bed bug bites all over my neck and arms—actual confirmed bites, unlike my mystery rash in London—it didn't bother me nearly as much as it would have a decade ago. I don't feel horror anymore, or even a deep worry that I'll bring any home. I followed my typical precautions in that Chicago hotel—I didn't put my luggage on the bed, I did laundry at the hotel once we discovered the bugs in the room, and when I got home I did laundry again and inspected my luggage.

Nearly four months later, as I worked through my manuscript revisions, there was still no sign that I brought any bed bugs home. And not only do I feel no fear or horror when I think about bed bugs (or spend the night with them), but I even have a begrudging admiration for the repellent little beasts. The bed bug is fascinating and the way it has both adapted to our lives and turned them upside down is quite a feat for such a small being. This bug is a survivor, and if you look at it from that angle, its story is impressive.

When I look at the vial of dead bed bugs that still sits on my desk—the gift from Harold Harlan—I'm reminded of a short story by the writer William Gass called "The Order of Insects," which he once said was the best he'd ever written. In it, a housewife describes mysterious roach carcasses that show up in her home each morning. She never sees them alive. At first, they horrify her, and she turns away as they are sucked up, shuddering as they rattle through the wand of the vacuum. But at some point, her perception shifts. She begins to examine the roaches' beautiful shells, the amazing geometric order of their hard bodies. And she starts collecting them in an old typewriter ribbon tin, occasionally opening it and marveling at the wonders inside. So, too, with my vial of New Jersey bugs, fed on the blood of a very curious scientist.

Let both curiosity and horror—the latter of which terrorizes us but also holds us rapt, unable to look away—be a motivator for discovery. Embrace the weird, the tiny, the things we'd like to ignore. Find joy in learning about them. In 1960 Karl von Frisch, the nineteenth-century ethologist and Nobel Laureate, wrote *Ten Little Housemates*, a slim book about the wonder that may be found in even the most reviled household pests. Of the bed bug, he wrote: "No animal is so insignificant that it might not reward our attention with great discoveries."

BROOKE'S BED BUG GUIDE

While writing this book, I was asked again and again for recommendations on how to deal with bed bugs. My response: I can't tell you what to do, because there are a lot of different methods that might work. But I will tell you what *I* would do in various situations. I make no promises that this will get rid of *your* bed bugs or prevent you from getting them. I also can't say that any of my preferences are right for a person in your specific living space or with your personal physical ailments or mental health issues. So don't sue me.

Would I hire a professional pest controller?

The experts will cringe at this, but if I thought I had bed bugs again, I would not call a pest control company immediately. This is partly because I am very allergic to bed bugs and would likely catch their presence early, before the situation reached emergency levels. Why wait? Treatment is expensive and the bugs are resistant to many products, while other options don't work at all. I would want to be really sure that I had bed bugs before handing hundreds of dollars or more over to an exterminator or forcing my landlady to do so. Repeatedly paying for bed bug treatment in a big city where the pest is common is simply not financially sustainable.

If I did find a bite and thought it happened while I slept in my bed, and I wasn't entirely sure it was from a bed bug, I would wait and see if I got more bites over the next few days or week. If I did, I'd do all my laundry, including my sheets and blankets, and I'd also give my bedroom a good cleaning and vacuuming. If I continued to get bites, I'd inspect my mattress, headboard, and box spring with a flashlight and look for bed bugs, shed exoskeletons, and the black speckles of bed bug feces. If I found any, and it was a very small patch, I would take whatever they were living on and either wash and dry it or spot heat-treat it (I have a steamer) or throw it out. I'd vacuum and possibly do laundry again. But if I continued to get bites after that, I'd admit defeat and call a professional.

Will my landlady pay for my bed bug treatment?

It depends. In New York, where I live, most owners are financially responsible for bed bug treatments, but it seems to vary with different types of buildings. If I got bed bugs again, I would research the legal options for my current building before approaching my landlady.

If I hired a pest controller, what chemicals would I make sure they use?

That's a tough one, and it will change over time. As I type these words, I would likely make sure my pest control operator used a pyrethroid-neonicotinoid combination, which currently seems to work best against bed bugs. But the insects are already proving resistant to this pesticide concoction, so it won't be the right choice forever. As for other chemicals, I personally wouldn't bother with insect growth regulators (IGRs) because they only appear to work at doses that are higher than can be legally applied. As for other options on the market, I'd probably read up on the latest academic research. Not sure where to look? Try searching for bed bugs and pesticides in databases such as Google Scholar, or look for articles on reputable science news sites.

What about those all-natural powders?

I assume you mean diatomaceous earth (DE), which is made of crushed-up fossilized diatoms, which are a kind of phytoplankton—tiny aquatic organisms. DE and similar powdered products are desiccants, and they basically dry up bed bugs and other insects and kill them. I would probably let my pest controller use DE, although recent research from the University of Kentucky calls its efficacy into question. But there isn't a big risk in applying it anyway, as long as you don't use such massive amounts that it hangs in clouds in the air. I have not found studies on the effect of DE on the lungs, but I doubt it would be good for your respiratory system. Also note that some desiccants are laced with pesticide, so make sure to read labels and follow instructions if you're using these products on your own.

What about those all-natural sprays?

As you can probably guess from my chapters that mention the minimum-risk 25b category of pesticide ingredients, I wouldn't bother with these or with any "contact killer." Their efficacy is questionable, and the Federal Trade Commission has gone after at least two companies that sell these products for false advertising. I wouldn't spend money on this type of product. Plus, I've grown skeptical of any product that uses "all-natural" as an advertising lure. This doesn't mean the product is good or safe, and it's an attempt to appeal to well-meaning people interested in sustainability and the environment.

But what about that new Neem oil that the EPA recently approved?

Here's what I know about Neem: I once saw a house in Ohio that had been treated with Neem for several months, and it still had bed bugs living in it. The product's smell also lingered on my coat for several hours after I left the house. It's a very strong scent. I probably wouldn't personally try it.

What about the dogs?

Dogs have been shown to be highly successful in locating bed bugs in a variety of tests, but only in very controlled laboratory settings. In unpublished field studies, the efficacy decreases significantly. If I had a pest controller bring a dog in, and that dog indicated I had

bed bugs, I would want a visual confirmation that the bugs actually existed before paying for a treatment.

What about heat treatments?

If I lived in a single-family home, had a really bad bed bug problem, and could afford it, I would opt for a heat treatment. But this
isn't the right choice for my Brooklyn apartment, because buildings with multiple units aren't easily treated with heat. I might,
however, let a company take my furniture and other belongings
and put them in a heat trailer specifically designed to kill bed bugs,
as long as they also treated my apartment with various other techniques, including pesticides, dusts, and so forth.

What about fumigants?

Same answer as for heat treatments.

What about traps?

I'm so allergic to bed bugs that if any got into my home, I'd probably notice the bites before the insects ended up in any traps. Because of this, I wouldn't bother with these, with one caveat. If I
lived somewhere that was really badly infested or had repeat infestations, I would consider both the traps that fit under bed legs and
some of the options with pheromone and CO_2 lures. These do seem
helpful to pest control operators or research entomologists who
are trying to confirm whether or not a treatment worked.

What about other over-the-counter products?

I wouldn't bother with most bed bug specialty products. The benefits seem more psychological than actual. I don't see the point in
spending hundreds of dollars on a self-heating suitcase that will
just get lost or smashed by an airline, or questionable sticky traps,
or various sprays. If I traveled more often—maybe more than once
a month—and stayed in hotels or hostels, I might invest in a small
self-heating box to treat my luggage each time I come home. I have

my own system for keeping bed bugs out of my house after traveling and it has worked well so far, at least since my original bed bug issues. I do laundry right away, as well as visually inspect my luggage and hit it with a steamer. This has seemed to work for the trips I've taken to bed-bug-infested public housing in Virginia and Ohio, as well as after spending nearly three weeks staying in hostels and hotels across Europe, although there's no way to be sure I simply was lucky and didn't actually encounter any bugs during these trips.

Any other tips for the home?

If I found out I had bed bugs, I would keep my clothes and extra bedding in sealed plastic bins and bags until I was sure the bugs were gone, even if this took weeks or even months (after washing/drying everything in high heat or steaming them, of course). I would also keep backpacks, purses, jackets, or other items that I regularly take out of my apartment and into crowded public spaces or other people's homes in a room far away from the bed, and regularly inspect them to make sure I am not spreading the problem to someone else. I might even run my steamer over my bag, jacket, and shoes before I leave the house, if the material could handle it.

What do I do when I stay in a hotel?

I never put my suitcase or bag on the bed. I keep my luggage by the front door when I first enter the room, and I check the mattress and headboard (if it isn't attached to the wall). This isn't a guarantee that I'll find bed bugs if they are there, but these are the most likely hiding places. If everything is clear, I will hang my clothes in the closet (but not touching the walls) and put my suitcase on a luggage rack (after inspecting it). I never put my clothes in the dresser drawers.

Some people go so far as to keep their luggage in the bathroom. That's a fine idea if you can handle a bathroom full of bags, especially if you're sharing your hotel room with another person. But I find it annoying.

Have I changed my behavior after writing this book?

Yes and no. I am more aware of my luggage when I travel, and I inspect it when I come home. I try to be more mindful of vacuuming. I still buy clothes at vintage shops, which is something I've always enjoyed, but I inspect them, seal them in a bag, and launder them as soon as I get a chance. I also still go to movies, libraries, and other public places—the chances of getting a bed bug in these environments are slim. When I go to parties and the host says, "Just throw your coat on the bed!" I make a joke and try to put it somewhere else. But even when I've followed their directions, I've never brought bed bugs home from a party. That I know of.

Overall, I think it is really important to just live your life. Sure, it can be trying to get bed bugs. But to let them control your decisions and keep you from doing things you enjoy—whether it's perusing secondhand stores, going to the theater, or something else entirely—seems, to me, a waste.

BED BUG SONGS

There are many bed bugs songs. This isn't all of them. But it's a lot of them.

"Honest" Abraham Allgood, "March of the Bedbugs" (2010)
Billy Banks, "Mean Old Bed Bug Blues" (1932)
Bill & Bink, "Bed Bug Boogie" (1958)
Andrew Bird, "Bed Bugs" (1998)
Blind Lemon Jefferson, "Black Snake Moan" (1927)
Blind Lemon Jefferson, "Chinch Bug Blues" (1927)
Brian & O'Brien, "Bed Bug Parody to Iron Maiden's 'Run to the Hills'"
    (2010)
David Bromberg, "Bedbug Blues" (1975)
Brooks, "Bedbugs" (2004)
Brownie with the Lad Richards Orchestra, "The Bedbug Song" (1961)
George Cahill, "Bed Bug Song" (2011)
Cat Cat Cat & Brother Nick, "The Bed Bug Song" (2011)
Caveman, "Bed Bugs" (2005)
Chizmo Charles, "Bed Bug Boogie" (1996)
The Cool Calm Collective, "Bed Bugs" (2011)
Crazy Spirit, "Bed Bugs" (2012)
Crooked Cowboy & the Freshwater Indians, "Bed Bugs" (2008)
De/Vision, "Bed Bugs (A Modern Lullaby)" (2012)
Echo and the Bunnymen, "Bedbugs and Ballyhoo" (1987)
Echo and the Bunnymen, "Bedbugs and Ballyhoo (Club Remix)" (1988)
Friday Night Flyers, "Bed Bugs" (2011)
The Gallop, "My Bed Bugs Fight" (2010)
The Galt Line, "Bedbugs" (2011)
Jerry Garcia and David Grisman, "Ain't No Bugs on Me" (1993)
Papa Joe Grappa, "Bedbug Boogie" (2010)
The Great Dictators, "Bed Bugs" (2012)
Hawkshaw Hawkins, "Mean Old Bed Bug Blues" (1946)
The Hubbards, "Bed Bugs" (2012)

Ill-Esha, "Bedbug" (2010)
The Johnnys, "Bedbug Banquet" (2010)
Lonnie Johnson, "Mean Old Bedbug Blues" (1927)
Laika, "Bed Bugs" (1997)
Kat Larios, "Bedbugs" (2008)
Lead Balloon, "Bedbugs" (2010)
Don Lennon, "Bedbugs" (2010)
Furry Lewis, "Mean Old Bedbug Blues" (1927)
Lightnin' Slim, "Bed Bug Blues" (1959)
The Limousines, "Bed Bugs" (2013)
Lord Invader, "Reincarnation" (1955)
Lord Kitchener, "Muriel and the Bug" (1961)
The Mighty Spoiler, "Bed Bug" (1940s)
Necro, "Bedbugs" (2010)
Noah23, "Bed Bugs" (2011)
The Pantookas, "Bedbugs" (1997)
Penpal, "Bed Bugs" (2011)
Picture Books, "Bed Bugs" (2009)
Poet & the Loops, "Return of the Bed Bugs London Olympics" (2011)
Pokey LaFarge & the South City Three, "Bed Bugs" (2011)
Pony Boy ft. Dagda and Paradime, "Bed Bugs" (2012)
The Rolling Stones, "Shattered" (1978)
Lee Ryan, "Bed Bugs" (2008)
Dmitri Shostakovich, "Incidental Music for *The Bedbug*" (1929)
Bessie Smith, "Mean Old Bedbug Blues" (1927)
Squirrel Nut Zippers, "Bedbugs" (2000)
Strange Arrangement, "Bed Bugs" (2011)
Summersett Fred, "Bed Bug" (2013)
Surgeon, "Bed Bugs" (2012)
Susquehanna Industrial Tool & Die Co., "Bed Bug Boogie" (2010)
Total Emergency, "Bed Bugs" (2009)
The Toy Dolls, "Bitten by a Bed Bug" (1991)
Ernest Tubb, "Mean Old Bed Bug Blues" (1936)
Antonia Vai, "Don't Let the Bedbugs Bite" (2010)
Dave Van Ronk, "Bed Bug Blues" (1991)
Fats Waller, "Mean Old Bed Bug Blues" (1930s)
Ignatz Zatz, "Bed Bugs Boogaloo" (2010)
Zawose & Brook, "Kuna Kunguni/The Bed Bugs Bite" (2001)

BED BUG LITERATURE

Bed bugs appear in many famous literary works. Here are brief excerpts from some of my favorites.

John Dos Passos, "Orient Express," in *Dos Passos: Travel Books & Other Writings 1916–1941* (New York: Literary Classics of the United States, 2003), p. 159.

> *By this time it was night. The train was joggling its desultory way through mountain passes under a sky solidly massed with stars like a field of daisies. In the crowded compartment, where people had taken off their boots and laid their heads on each other's shoulders to sleep, hordes of bedbugs had come out of the stripped seats and bunks, marching in columns of three or four, well disciplined and eager. I had already put a newspaper down and sprinkled insect powder in the corner of the upper berth in which I was hemmed by a solid mass of sleepers. The bedbugs took the insect powder like snuff and found it very stimulating, but it got into my nose and burned, got into my eyes and blinded me, got into my throat and choked me, until the only thing for it was to climb into the baggage rack, which fortunately is very large and strong in the Brobdignagian Russian trains. There I hung, eaten only by the more acrobatic of the bugs, the rail cutting into my back, the insect powder poisoning every breath, trying to make myself believe that a roving life was the life for me. Above my head I could hear the people on the roof stirring about.*

Allen Ginsberg, "Death on All Fronts: 'The Planet Is Finished,'" *The Fall of America: Poems of These States 1965–1971* (San Francisco: City Lights Books, 1972), p. 131.

> *. . . I called in Exterminator Who soaked the Wall floor with*
> *bed-bug death-oil. Who'll soak my brain with death-oil?*

Johann Wolfgang von Goethe, *Works of Johann von Goethe*, trans. Bayard Taylor (Cambridge: Riverside Press, 1883), p. 54.

> *The lord of rats and eke of mice*
> *Of flies and bed-bugs, frogs and lice,*

> *Summons thee hither to the door-sill,*
> *To gnaw it where, with just a morsel*
> *Of oil, he paints the spot for thee:—*
> *There com'st thou, hopping on to me!*
> *To work, at once! The point which made me craven*
> *Is forward, on the ledge, engraven.*
> *Another bite makes free the door:*
> *So, dream thy dreams, O Faust, until we meet once more!*

Horace, "Poetic Cliques," in *Horace Talks*, trans. Henry Harmon Chamberlin (Hallandale, FL: New World Book Manufacturing, 1940), p. 101.

> *Pantilius, the bed bug, shall he sting*
> *Me, or Demetrius my withers ring*
> *Talking against me when I am not present?*

Langston Hughes, *Not without Laughter* (Mineola: Dover, 2008), p. 285.

> *Upstairs he found his mother sleeping deeply on one side of the bed. He undressed, keeping his underwear, and crawled in on the other side, but he lay awake a long while because it was suffocatingly hot, and very close in their room. The bed-bugs bit him on the legs. Every time he got half asleep, an L train roared by, shrieking outside their own windows, light up the room, and shaking the whole house.*

Aldous Huxley, *Antic Hay* (London: Dalkey Archive Press, 2006), p. 55.

> *"What indeed?" said Coleman. "Tics, mere tics. Sheep ticks, horse ticks, bed bugs, tape worms, taint worms, guinea worms, liver flukes . . ."*

David Herbert Lawrence, "Proper Pride," *The Complete Poems of D. H. Lawrence* (Hertfordshire: Wordsworth Editions Limited, 1994), p. 514.

> *Everything that lives has its own proper pride*
> *as a columbine flower has, or even a starling walking and looking*
> *around.*
> *And the base things like hyenas or bed-bugs have least pride of*
> *being,*
> *they are humble, with a creeping humility, being parasites or carrion*
> *creatures.*

Sinclair Lewis, *Elmer Gantry* (New York: New American Library, 1970), p. 86.

> *The walls were of old plaster, cracked and turned deathly gray, marked with the blood of mosquitoes and bed-bugs slain in portentous battles long ago by theologians now gone forth to bestow their thus uplifted visions on a materialist world.*

Henry Miller, *Black Spring* (New York: Grove Press, 1963), p. 39.

> *"I accept Time absolutely." Whereas my friend Carl, who has the vitality of a bedbug, is pissing in his pants because four days have elapsed and he only has a negative in his hand. "I don't see any reason," says he, "why I should ever die—barring an untoward accident." And then he rubs his hands and closets himself in his room to live out his immortality. He lives on like a bedbug hidden in the wallpaper.*

Henry Miller, *Moloch; or, This Gentile World* (New York: Grove Press, 1992), p. 6.

> *Her consort was depicted stealing about the premises with a squirt gun. The imbecilic glee he displayed was apparently evoked by the sight of a filthy mattress from which an interminable file of bedbugs issued. (The bedbug is known to scientists as Cimex lectularius: a cosmopolitan blood-sucking wingless depressed bug of reddish brown color and vile odor, infesting houses and especially beds. The cockroach is the natural enemy of the bedbug.) Even the counterpane on which the assassin's saffron paramour reclined, after the now classic manner of Olympe, was diapered with these cosmopolitan blood-sucking wingless depressed bugs of reddish brown color and vile odor.*

Henry Miller, *Nexus* (New York: Grove Press, 1965), p. 42.

> *Peaceful as a bedbug I slept. Shortly after dawn I opened my eyes, amazed to discover I was not in the great beyond. Yet I could hardly say that I was still among the living. What had died I know not. I know only this, that everything which serves to make what is called "one's life" had faded away.*

Henry Miller, *Plexus* (New York: Grove Press, 1965), pp. 356, 357.

> *Mice, ants, cockroaches, bedbugs, every sort of vermin infested the place. The tables, beds, chairs, divans, commodes were littered with papers, with open file boxes, with cards, graphs, statistical tables, instruments of all kinds.*
>
> ...
>
> *We rose early, unable to sleep because of the bedbugs. We took a quick shower, examined our clothes thoroughly to make sure they were not infested, and prepared to decamp.*

Henry Miller, *Sexus* (New York: Grove Press, 1965), p. 405.

> *It was all going like tick-tack-toe, one thing canceling another, and at the end of course the law squashing you down as if you were a fat, juicy bedbug, when suddenly I realized that he was asking me if I were willing to pay such and such an amount of alimony regularly for the rest of my days.*

Henry Miller, *Tropic of Cancer* (New York: Grove Press, 1961), p. 152.

> *That's the first thing that strikes an American woman about Europe—that it's unsanitary. Impossible for them to conceive of a paradise without mod-*

ern plumbing. If they find a bedbug they want to write a letter immediately to the chamber of commerce. How am I ever going to explain to her that I'm contented here? She'll say I've become a degenerate.

Henry Miller, *Tropic of Capricorn* (New York: Grove Press, 1961), p. 319.

The hibernation of animals, the suspension of life practiced by certain low forms of life, the marvelous vitality of the bedbug which lies in wait endlessly behind the wallpaper, the trance of the Yogi, the catalepsy of the pathologic individual, the mystic's union with the cosmos, the immortality of cellular life, all these things the artist learns in order to awaken the world at the propitious moment.

Evelyn Waugh, *Brideshead Revisited* (Boston: Back Bay Books/Little, Brown, 1944), p. 155.

When he came it was, of course, with perfect propriety; he apologized, sat in the empty place and allowed Mr. Samgrass to resume his monologue, uninterrupted and, it seemed, unheard. Druses, patriarchs, icons, bed-bugs, romanesque remains, curious dishes of goat and sheeps' eyes, French and Turkish officials—all the catalogue of Near Eastern travel was provided for our amusement.

John Steinbeck, *Tortilla Flat* (New York: Penguin Books, 2000), pp. 8.

The bedbugs bothered him a little at first, but as they got used to the taste of him and he grew accustomed to their bites, they got along peacefully.

He started playing a satiric game. He caught a bedbug, squashed it against the wall, drew a circle around it with a pencil and named it "Mayor Clough." Then he caught others and named them after the City Council. In a little while he had one wall decorated with squashed bedbugs, each named for a local dignitary. He drew ears and tails on them, gave them big noses and mustaches. Tito Ralph, the jailer, was scandalized; but he made no complaint because Danny had not included either the justice of the peace who had sentenced him, nor any of the police force. He had a vast respect for the law.

Upton Sinclair, *The Jungle* (Upton Sinclair, 1920), pp. 188–89.

There were two bunks, one above the other, each with a straw mattress and a pair of gray blankets—the latter stiff as boards with filth, and alive with fleas, bed-bugs, and lice. When Jurgis lifted up the mattress he discovered beneath it a layer of scurrying roaches, almost as badly frightened as himself.

BED BUG LIMERICKS

I wrote three bed bug limericks. I should probably be embarrassed about this, but instead I share them with you here:

Limerick 1

There once was a C. lectularius
Whose odor was very nefarious
He smelled like old fruit
So foul and acute
So nasty, it was quite hilarious

Limerick 2

There once was a bed bug named Peter
Whose bite was much worse than a skeeter's
He'd suck up your blood
Just open a flood
By the end, he'd drain a whole liter

Limerick 3

The bed bug traumatically inseminates
Much to the dismay of his many mates
Not one for romance
He stabs with his lance
No surprise he's not asked on many dates

ACKNOWLEDGMENTS

While dreaming of writing a book isn't unusual in itself, it may be if the subject makes people writhe and squirm. For having countless conversations about bed bugs, as well as for endless encouragement, I would first like to thank my husband, Mike Wasilewski, and both of our immediate and extended families and friends. You have all put up with a tremendous amount of information on bed bugs over a period of several years with such gusto, and for that I am grateful.

I would also like to thank my agent, Paul Lucas, and the other fine people at Janklow & Nesbit Associates for seeing potential in my idea and supporting me in pursuing it. Thank you as well to my editor, Christie Henry—especially for the hand-holding I required as a first-time author—as well as Carrie Adams, Erin DeWitt, Amy Krynak, and everyone else at the University of Chicago Press for their lovely work.

This book would not be possible without the additional support I received from the Alfred P. Sloan Foundation; thanks go to Doron Weber, Delia DiBiasi, Sonia Epstein, and all of the other folks at that laudable institution. An enormous thank-you also goes to everyone who donated to the Kickstarter campaign I launched early on in this project, and especially to Chris Handy, Martha Harbison, Mike Jacobellis, Ben Lillie, Karl Korinek, Astin Little, Lori Marino, Erin Mindell, Jennifer Nadeau, Brian Ogilvie, Eric Pedersen, and Zachary Yorke.

There are so many more people to thank: Seth Fletcher for his insight on book proposals, Jessica Seigel for the title inspiration, Emily Voigt for our invaluable two-person writing group at Good Stuff Diner, and Mary Roach and Jonny Waldman for their advice and encouragement; the artists and photographers who have allowed me to showcase their work; my readers who graciously gave their time and wisdom on early drafts, including Adrianna Dufay, Martha Harbison, Dan Vergano, Natalie Wolchover, and the review board from the University of Chicago Press; and my dogged proofreaders, Julia Calderone and Matt Koreiwo. I would also like to give special thanks to Stephen Doggett, who read a version of this book for entomological accuracy, although I take full responsibility for all of the facts, explanations, and mistakes in the final product. A big thank-you to those who helped me dig up references and images, including the Entomological Society of America and the US Armed Forces Pest Management Board, as well as the researchers and librarians at the Brooklyn Public Library, the New York Public Library, the Crossett Library at Bennington College, the Lewis Walpole Library at Yale University, the London School of Hygiene and Tropical Medicine, the Camden Local Studies and Archives Centre, and the Essig Museum of Entomology and the Bancroft Library, both at the University of California, Berkeley (with special thanks to Maria Brandt for digging through Bancroft boxes and meticulously photographing the contents so I could read them across the country).

Finally, I would like to thank all of my sources, both those whose work and stories ended up in the book and those whose did not. You were all so generous with your time and your knowledge, without which this project wouldn't have been possible regardless of the tremendous network I've listed above.

Thank you, thank you, thank you.

REFERENCES

EPIGRAPHS
The three epigraphs in the book's front matter come from the following sources, which are listed in the order in which the quotes appear.

1. L. Howard, *The Insect Book: A Popular Account of the Bees, Wasps, Ants Grasshoppers, Flies and Other North American Insects Exclusive of the Butterflies Moths and Beetles, with Full Life Histories, Tables and Bibliographies* (New York: Doubleday, Page & Co., 1902).
2. F. Brown, *The Frank C. Brown Collection of NC Folklore*. Vol. V: *The Music of the Folk Songs* (Durham, NC: Duke University Press, 1962); D. Dance, *From My People: 400 Years of African American Folklore: An Anthology* (New York: Norton, 2002); A. Bontemps, "Bed Bug," *An Anthology of African American Poetry for Young People* (Smithsonian Folkway Recordings, 1990).
3. C. Taylor, "Unwelcome Guests," *Harper's New Monthly Magazine*, December 1860.

PROLOGUE
Statistics come from the following sources, listed in which the order they appear.

1. M. Singer, "Dept. of Entomology: Night Visitors," *New Yorker*, April 4, 2005.
2. New York Department of Housing Preservation and Development, Bed Bug Complaints and Violations from 2004 to 2010.
3. National Pest Management Association, "Bugs without Borders Survey," 2011.
4. National Pest Management Association, "Bugs without Borders Survey," 2013.
5. S. Doggett and R. Russell, "The Resurgence of Bed Bugs, *Cimex* spp. (Hemiptera: Cimicidae) in Australia: Experiences from Down Under," *Proceedings of the Sixth International Conference on Urban Pests*, ed. W. H. Robinson and D. Bajomi (Budapest: Executive Committee of the International Conference of Urban Pests, 2008).

CHAPTER ONE
Sources for this chapter are listed below, in the order in which they appear. Sources used multiple times, in particular *Monograph of Cimicidae* by Robert Leslie Usinger, are listed only once.

1. S. Doggett et al., "Bed Bugs: Clinical Relevance and Control Options," *Clinical Microbiology Reviews* 25, no. 1 (2012): 164–92.

2. L. Reinhardt and M. Siva-Jothy, "Biology of the Bed Bug (Cimicidae)," *Annual Review of Entomology* 52 (2007): 351–74.

3. R. L. Usinger, *Monograph of Cimicidae* (Lanham, MD: Entomological Society of America, 1966).

4. J. Goddard, "Effects of Bed Bug (*C. lectularius* L.) Saliva on Human Skin" (presentation at the annual meeting of the Entomology Society of America, Knoxville, Tennessee, November 11–14, 2012).

5. G. Arnqvist and L. Rowe, *Sexual Conflict: Monographs in Behavior and Ecology* (Princeton, NJ: Princeton University Press, 2005).

6. I. Francischetti et al., "Insight into the Sialome of the Bed Bug, *Cimex lectularius*," *Journal of Proteome Research* 9 (2010): 3820–31.

7. M. Lehane, *The Biology of Blood-Sucking in Insects*, 2nd ed. (Cambridge: Cambridge University Press, 2005).

8. F. Ko et al., "Engineering Properties of Spider Silk," *MRS Proceedings* 702 (2002).

9. R. Fogain, *Beat Bed Bugs and Other Pests: Learn How to Rid Your House of the Critters* (Victoria, BC: Friesen Press, 2013).

10. A. Stutt and M. Siva-Jothy, "Traumatic Insemination and Sexual Conflict in the Bed Bug *Cimex lectularius*," *PNAS* 98, no. 10 (2001): 5683–87.

11. R. Naylor, interview with author, July 17, 2012.

12. R. Green et al., "A Draft Sequence of the Neanderthal Genome," *Science* 328, no. 5979 (2010): 710–22.

13. O. Balvin et al., "Mitochondrial DNA and Morphology Show Independent Evolutionary Histories of Bedbug *Cimex lectularius* (Heteroptera: Cimicidae) on Bats and Humans," *Parasitology Research*, March 6, 2012.

14. R. Dunn, "Bedbugs Have Evolved to Live with Mankind," *Miller-McCune*, March 24, 2011.

15. E. Panagiotakopulu and P. Buckland, "*Cimex lectularius* L., the Common Bed Bug from Pharaonic Egypt," *Antiquity* 73 (1999): 908–11.

16. E. Strouhal, *Life of the Ancient Egyptians* (Norman: University of Oklahoma Press, 1992).

17. J. Busvine, *Insects, Hygiene and History* (London: Athlone Press, 1976).

18. E. Butler, *Our Household Insects: An Account of the Insect-Pests Found in Dwelling-Houses* (London: Longmans, Green, and Co., 1893).

19. J. K. Elliot, *The Apocryphal New Testament* (New York: Oxford University Press, 2005).

20. M. Vidas, e-mail message to author, March 23, 2012.

21. P. Lucas, e-mail message to author, November 8, 2012.

22. E. Perlstein, Facebook message to author, November 8, 2012.

23. D. Varisco, e-mail message to author, April 10, 2012.

24. D. Powers, e-mail message to author, April 12, 2012.

25. G. De Cuvier, *The Animal Kingdom, Arranged According to Its Organization; Forming the Basis for a Natural History of Animals, and an Introduction to Comparative Anatomy* (London: Wm. S. Orr and Co., 1840).

26. B. Cummings, *The Bed-Bug: Its Habits and Life History, and How to Deal with It* (London: British Museum, 1949).

27. J. Kieran, *A Natural History of New York City* (New York: Houghton Mifflin, 1959).

28. P. Kalm, *Kalm's Account of His Visit to England on His Way to American in 1748* (London: Macmillan, 1892).

29. P. Kalm, *Travels into North America* (London, 1771).

30. J. Stratton and A. Schlesinger, *Pioneer Women: Voices from the Kansas Frontier* (New York: Simon & Schuster, 1981).

31. D. Gray, *Women of the West* (Nebraska: Dorothy Kamer Gray, 1976).

32. S. Doggett and R. Russell, "The Resurgence of Bed Bugs, *Cimex* spp. (Hemiptera: Cimicidae) in Australia: Experiences from Down Under," *Proceedings of the Sixth International Conference on Urban Pests*, ed. W. H. Robinson and D. Bajomi (Budapest: Executive Committee of the International Conference of Urban Pests, 2008).

33. S. Doggett, e-mail message to author, April 9, 2012.

34. A. Girault, "Notes on the Feeding Habits of *Cimex lectularius*," *Psyche* 15, no. 4 (1908): 85–87.

35. R. Overstreet, "Flavor Buds and Other Delights," *Journal of Parasitology* 89, no. 6 (2003): 1093–107.

36. R. Dunglison, *Medical Lexicon: A New Dictionary of Medical Science* (Philadelphia: Lea and Blanchard, 1839).

37. J. Clarke, *A Dictionary of Practical Materia Medica*, 3 vols. (1921; reprint, New Delhi: B. Jain Publishers, 2006).

38. G. Levin, *Edward Hopper: An Intimate Biography* (Berkeley: University of California Press, 1995).

39. E. McDaniel, "Fleas and Bed-Bugs," *Michigan Circular Bulletin No. 94*, Michigan State University, June 1926.

40. Japanese translations of "bed bug" came from responses to an inquiry by the author posted to the H-Japan academic listserv, April 16, 2012.

41. A. Neundorf, *A Navajo/English Bilingual Dictionary* (Pine Hill, NM: Native American Materials Development Center, 1983).

42. C. Carroll, e-mail message to author, November 13, 2012.

43. E. Malotki, e-mail message to author, March 27, 2012.

44. F. Wan and F. Fangyu Wang, *Mandarin Chinese Dictionary: English-Chinese* (South Orange, NJ: Seton Hall, 1971).

45. K. Ch'ien, e-mail message to author, November 6, 2012.

46. H. Chamberlin, *Horace Talks* (Hallandale, FL: New World Book Manufacturing, 1940).

47. J. W. von Goethe, *Works of Johann von Goethe*, trans. Bayard Taylor (Cambridge: Riverside Press, 1883).

48. U. Sinclair, *The Jungle* (Upton Sinclair, 1920).

49. A. Huxley, *Antic Hay* (London: Dalkey Archive Press, 2006).

50. S. Lewis, *Elmer Gantry* (New York: New American Library 1970).

51. L. Hughes, *Not without Laughter* (Mineola, NY: Dover, 2008).

52. E. Goodman, *Writing the Rails* (New York: Black Dog & Leventhal, 2001).

53. W. S. Burroughs, *Naked Lunch* (New York: Grove Press, 2001).

54. A. Ginsberg, *The Fall of America: Poems of These States 1965–1971* (San Francisco: City Lights Books, 1972).

55. J. Dos Passos, "Orient Express" in *Dos Passos, in Travel Books & Other Writings 1916–1941* (New York: Literary Classics of the United States, 2003).
56. H. Miller, *Tropic of Cancer* (New York: Grove Press, 1961).
57. H. Miller, *Tropic of Capricorn* (New York: Grove Press, 1961).
58. H. Miller, *Black Spring* (New York: Grove Press, 1963).
59. H. Miller, *Sexus* (New York: Grove Press, 1965).
60. H. Miller, *Plexus* (New York: Grove Press, 1965).
61. H. Miller, *Nexus* (New York: Grove Press, 1965).
62. H. Miller, *Moloch; or, This Gentile World* (New York: Grove Press, 1992).
63. J. Steinbeck, *Tortilla Flat* (New York: Penguin, 1977).

Some of the passages of the chapter were adapted from a piece previously published by the author: portions of the description of bed bug sex appeared in a guest post "TGIPF: The Bed Bug and His Violent Penis," *The Last Word on Nothing*, July 27, 2012, part of the blog's "Thank God It's Penis Friday" series.

CHAPTER TWO
Sources for this chapter are listed below, in the order in which they appear. Sources used multiple times are listed only once.

1. F. Augustin, "Zur Geschichte des Insektizids Dichlordiphenyltrichlorä than (DDT) unter besonderer Berücksichtigung der Leistung des Chemikers Paul Müller (1899–1965)" (diss., Leipzig University).
2. "Obituary: Dr. Paul Müller," *Nature* 208 (1965): 1043–44.
3. *Nobel Lectures in Physiology or Medicine: 1942–1962* (Singapore: World Scientific Publishing Co. Pte. Ltd., 1999).
4. R. L. Usinger, *Monograph of Cimicidae* (Lanham, MD: Entomological Society of America, 1966).
5. "Paul Hermann Müller," Nobelprize.org, http://www.nobelprize.org/nobel _prizes/medicine/laureates/1948/muller-bio.html.
6. F. Leroy, *A Century of Nobel Prize Recipients: Chemistry, Physics, and Medicine* (New York: Marcel Dekker, 2003).
7. P. Ellis and A. MacDonald, *Reading into Science: Biology* (Cheltenham, UK: Nelson Thomas Ltd., 2003).
8. R. MacPherson, "A Modern Approach to Pest Control," *Canadian Journal of Comparative Medicine* 11, no. 4 (1947): 108–13.
9. J. Harbster, "World War II 'Scientific Manpower,'" *Inside Adams, Science, Technology, and Business*, November 10, 2010.
10. F. D. Roosevelt, "Executive Order 8807 Establishing the Office of Scientific Research and Development," in Gerhard Peters and John T. Woolley, *The American Presidency Project*, June 28, 1941.
11. "DDT," *Time*, June 12, 1944.
12. "Super-Delouser," *Newsweek*, June 12, 1944.
13. H. Zinsser, *Rats, Lice and History* (New Brunswick, NJ: Transaction, 2008).
14. H. Zinsser and M. Castaneda, "Active and Passive Immunization in Typhus Fever," *PNAS* 20, no. 1 (1934): 9–11.
15. *Encyclopedia of Pestilence, Pandemics, and Plagues*, ed. Joseph Patrick Byrne (Westport, CT: Greenwood Publishing, 2008).

16. K. Raper, "A Decade of Antibiotics in America," *Mycologia* 44, no. 1 (1952): 1–59.
17. S. Meshnick and M. Dobson, "The History of Antimalarial Drugs," *Antimalarial Chemotherapy: Mechanisms of Action, Resistance, and New Directions in Drug Discovery*, ed. P. J. Rosenthal (Totowa, NJ: Humana Press, 2001).
18. Paul Müller, Nobel Prize in Physiology or Medicine, 1948, award ceremony speech, http://www.nobelprize.org/nobel_prizes/medicine/laureates/1948/press.html.
19. P. Russell, *Preventative Medicine in World War II*, vol. 7: *Communicable Diseases, Malaria* (Washington, DC: United States Army Medical Service, 1955).
20. M. Blumenson, *The Patton Papers, 1940–1945* (Boston: Da Capo Press, 1974).
21. G. Sumner Jr., *Marching On: A General's Tales of War and Diplomacy* (Oakland, OR: Red Anvil Press, 2004).
22. D. Kinkela, *DDT and the American Century: Global Health, Environmental Politics, and the Pesticide that Changed the World* (Chapel Hill: University of North Carolina Press, 2011).
23. D. Stapleton, "The Short-Lived Miracle of DDT," *American Heritage of Invention and Technology* 15, no. 3 (2000): 34–41.
24. C. Wheeler, "Control of Typhus in Italy 1943–1944 by Use of DDT," *American Journal of Public Health* 36 (February 1946): 119–29.
25. F. Soper et al., "Typhus Fever in Italy, 1943–1945, and Its Control with Louse Powder," *American Journal of Hygiene* 45, no. 3 (1947): 305–34.
26. Armed Forces Pest Management Board, "US Army Entomology: A Historical Perspective," http://www.afpmb.org/content/united-states-army-medical-entomology.
27. M. Smallman-Raynor and A. Cliff, "Impact of Infectious Diseases on War," *Infectious Disease Clinics of North America* 18 (2004): 341–68.
28. T. Dunlap, *DDT: Scientists, Citizens, and Public Policy* (Princeton, NJ: Princeton University Press, 1981).
29. Environmental Protection Agency, "DDT Ban Takes Effect," http://www2.epa.gov/aboutepa/ddt-ban-takes-effect.
30. G. Ware and D. Witacre, *The Pesticide Book* (Willoughby, OH: MeisterPro Information Resources, 2004).
31. H. Greim and R. Snyder, *Toxicology and Risk Assessment: A Comprehensive Introduction* (West Sussex: John Wiley & Sons, 2008).
32. C. Boase, interviews with author, September 11, 2012; January 31, 2013.
33. K. Reinhardt, interviews with author, September 6–7, 2012.
34. L. Pinto et al., *Bed Bug Handbook* (Mechanicsville, MD: Pinto & Associates, 2007).
35. D. Biehler, *Pests in the City: Flies, Bedbugs, Cockroaches, & Rats* (Seattle: University of Washington Press, 2013).

CHAPTER THREE

Sources for this chapter are listed below, in the order in which they appear. Sources used multiple times are listed only once.

1. C. Barnhart, "The Use of Arthropods as Personnel Detectors," US Army Limited War Laboratory Aberdeen Proving Ground, Maryland, August 9, 1968.

2.   W. Beecher, "Bedbug May Help to Hunt Vietcong," *New York Times*, June 6, 1966.

3.   E. Hymoff, "Stalemate in Indo-China, Technology vs. Guerillas," *Bulletin of the Atomic Scientists*, November 1971.

4.   Analysis of bed bug research trends based on database, compiled by the author, containing research articles available in major databases (PubMed, Wiley, SciDirect, Springer, Taylor & Francis, PLOS ONE, and a private list from USDA entomologist Mark Feldlaufer) using the search terms "bed bug" OR bedbug OR "bed bugs" OR bedbugs or "cimex lectularius." Databases maintained and analyzed in Microsoft Excel).

5.   R. L. Usinger, *Monograph of Cimicidae* (Lanham, MD: Entomological Society of America, 1966).

6.   A. Brown, "The Insecticide-Resistance Problem: A Review of Developments in 1956 and 1957," *Bulletin of the World Health Organization* 18 (1958): 309–21.

7.   D. Micks, "Insecticide-Resistance: A Review of Developments in 1958 and 1959," *Bulletin of the World Health Organization* 22 (1960): 519–22.

8.   J. Busvine, "Insecticide-Resistance in Bed-Bugs," *Bulletin of the World Health Organization* 19 (1958): 1041–52.

9.   N. Gratz, "A Survey of Bed-Bug Resistance to Insecticides in Israel," *Bulletin of the World Health Organization* 20 (1959): 835–40.

10. J. Reid, "Resistance to Dieldrin and DDT and Sensitivity to Malathion in the Bed-bug *Cimex hemipterus* in Malay," *Bulletin of the World Health Organization* 22 (1960): 586–87.

11. J. Busvine, "Resistance to Pyrethrins," *Bulletin of the World Health Organization* 22 (1960): 592–93.

12. N. Gratz, "Insecticide-Resistance in Bed-bugs and Flies in Zanzibar," *Bulletin of the World Health Organization* 20 (1959): 835.

13. A. Brown, "Insecticide Resistance and the Future Control of Insects," *Canadian Medical Association Journal* 100 (1969): 216–21.

14. R. L. Usinger, *Robert Leslie Usinger: Autobiography of an Entomologist* (San Francisco, CA: Pacific Coast Entomological Society, California Academy of Sciences, 1972).

15. E. Ross, interview with author, March 29, 2013.

16. R. Usinger, interviews with author, February 21, 2013; March 6, 2013.

17. E. G. Linsley, "Robert Leslie Usinger: 1912–1968," *Pan-Pacific Entomologist* 45 (1969): 167–84.

18. D. Heyneman, interview with author, March 21, 2013.

19. Various correspondence from 1957 to 1968, Povolný folder, box 2, Usinger Papers, Bancroft Library, University of California, Berkeley.

20. H. Harlan, interview with author, July 8, 2011.

21. H. Harlan, e-mail message to author, July 22, 2011.

22. H. Harlan, "My Bed Bug Population's History" (unpublished manuscript, September 2010).

23. J. Bartley and H. Harlan, "Bed Bug Infestation: Its Control and Management," *Military Medicine* 39, no. 11 (1974): 884–86.

24. D. Kinkela, *DDT and the American Century: Global Health, Environmental Politics, and the Pesticide that Changed the World* (Chapel Hill: University of North Carolina Press, 2011).

25. R. Carson, *Silent Spring* (New York: Houghton Mifflin, 2002).

26. W. Sladen et al., "DDT Residues in Adelie Penguins and A Crabeater Seal from Antarctica," *Nature* 210 (1966): 670–73.

27. National Occanic and Atmospheric Administration, "Persistent Organic Pollutants in the Arctic," *Arctic Pollution Issues: A State of the Arctic Environment Report*, http://www.arctic.noaa.gov/essay_calder.html.

28. T. Dunlap, *DDT: Scientists, Citizens, and Public Policy* (Princeton, NJ: Princeton University Press, 1981).

29. "DDT—A Brief History and Status," last modified May 9, 2012, http://www.cpa.gov/oppooool/factsheets/chemicals/ddt-brief-history-status.htm.

30. "DDT Regulatory History: A Brief Survey (to 1975)," http://www2.epa.gov/aboutepa/ddt-regulatory-history-brief-survey-1975.

31. "EPA History: Federal Insecticide, Fungicide and Rodenticide Act," http://www2.epa.gov/aboutepa/epa-history-federal-insecticide-fungicide-and-rodenticide-act.

32. "Montrose Chemical Company," http://yosemite.epa.gov/r9/sfund/r9sfdocw.nsf/vwsoalphabetic/Montrose+Chemical+Corp?OpenDocument.

33. H. Rafatjah, "The Problem of Resurgent Bed-bug Infestation in Malaria Eradication Programs," *Journal of Tropical Medical Hygiene* 74, no. 2 (1971): 53–56.

CHAPTER FOUR

Sources for this chapter are listed below, in the order in which they appear. Sources used multiple times are listed only once.

1. P. Leschen and F. Sauter, *BEDBUGS!!! The Comedy Sci-fi Thriller Rock Musical*, ATA's Chernuchin Theatre, November 3, 2012.

2. P. Heckman, e-mail message to author, November 11, 2013.

3. F. Sauter, e-mail message to author, November 12, 2013.

4. Broadway World News Desk, "Pit Stop Players Play DiMenna Center 5/10," Broadwayworld.com, April 12, 2011, http://www.broadwayworld.com/article/Pit-Stop-Players-Play-DiMenna-Center-510-20110412.

5. M. Deterior et al., *Save the Bed Bugs*, issues 1–4 (2011–2013).

6. S. Oliver, *The Exterminators: Bug Brothers* (New York: DC Comics, 2006).

7. S. Oliver, *The Exterminators: Crossfire and Collateral* (New York: DC Comics, 2007).

8. J. Salicrup, *The Three Stooges: Bed-Bugged!* (New York: Papercutz, 2012).

9. B. Winters, *Bedbugs* (Philadelphia: Quirk Books, 2011).

10. B. Winters, interview with author, February 6, 2013.

11. R. Schapiro, "Bedbugs Found in Small Area in Basement of Empire State Building," *New York Daily News*, August 21, 2010.

12. Post Staff Report, "Bedbugs Continue Assault on NYC at the U.N.," *New York Post*, October 27, 2010.

13. A. Hunter, "Bed Bug Lingerie Infestation: Really Gross, but Health Hazard?" *CBS News*, July 20, 2010.

14. M. West, "Waldorf Astoria Blamed for Family's Bedbugs," *Wall Street Journal*, October 21, 2010.

15. J. Doll, "Are Bed Bugs Biting at the New York Public Library?" *Village Voice*, August 5, 2010.

16. D. Anderson, "Bedbugs in Flight," *New York Times*, August 16, 2010.

17. G. Marush, "Bed Bugs Found in Five Residence Halls," *GW Hatchet*, November 18, 2010.

18. "Grove City Motel Cited for Bed Bugs," 10TV.com, September 13, 2010.

19. CBS, "2010: Year of the Bed Bug?" December 21, 2010, http://www.cbsnews.com/video/watch/?id=7172145n.

20. C. Longrigg, "Voracious Bed Bugs Put Bite on Budget Hotel Tourists," *Guardian*, September 8, 1997.

21. "Bedbugs Are Biting Back," *Daily Telegraph*, April 22, 2000.

22. Search via the *New York Times* website with terms "bed bug" OR bedbug, February 24, 2013.

23. Search via the INFOTRAC Newsstand database, with terms "bed bug" OR bedbug OR "bed bugs" OR bedbugs OR "Cimex lectularius," searched February 27, 2013.

24. J. Howell, "A Guide to Georgia Insects: You Won't Sleep Easy with Bedbugs Around," *Atlanta Journal-Constitution*, February 14, 2002.

25. J. Greig, "Don't Let the Bedbugs Bite (or Overwhelm Your House)," *Austin American-Statesman*, October 24, 2003.

26. R. Fagerlund, "The Bug Man: Bed Bugs Are Irritating but Not Harmful," *Santa Fe New Mexican*, March 9, 2003.

27. M. May, "Bedbugs Bounce Back: Outbreaks in All 50 States," *San Francisco Chronicle*, April 8, 2007.

28. K. Brumback, "Bedbug Infestations Are on the Rise, Experts Say," *Cincinnati Post*, August 8, 2006.

29. L. Lapeter, "They Bite, and They're Back: Bedbugs. Really," *St. Petersburg Times*, July 3, 2005.

30. M. Singer, "Night Visitors," *New Yorker* April 4, 2005.

31. M. Davey, "Step Right Up for the Pest Control at Bedbug Meeting," *New York Times*, September 21, 2010.

32. T. Russel, "Alone When the Bedbugs Bite," *New York Times*, November 18. 2010.

33. C. Saint Louis, "A Dark and Itchy Night," *New York Times*, December 5, 2012.

34. J. Eisenberg, interview with author, January 8, 2013.

35. C. Boase, interview with author, January 31, 2013.

36. T. St. Amand, interview with author, September 15, 2011.

37. R. Cooper, interview with author, January 31, 2013.

38. R. Vetter and G. Isbister, "Medical Aspects of Spider Bites," *Annual Review of Entomology* 53 (2008): 409–29.

39. R. Vetter, e-mail message to author, January 24, 2013.

40. R. Vetter, interview with author, August 13, 2012.

41. L. Pinto et al., *Bed Bug Handbook* (Mechanicsville, MD: Pinto & Associates, 2007).

42. Environmental Protection Agency, "Pyrethroids and Pyrethrins," last updated October 2012, http://www.epa.gov/oppsrrd1/reevaluation/pyrethroids-pyrethrins.html.

43. R. Cooper, interview with author, January 31, 2013.

44. R. Cooper, e-mail message to author, February 26, 2013.

45. D. Miller, interviews with author, April 18–19, 2013.

46. M. Potter, interviews with author, November 20, 2012; January 28, 2013; February 12, 2013.

47. M. Potter, various e-mail messages to author, June 6, 2011–January 13, 2014.

48. D. Biehler, *Pests in the City: Flies, Bedbugs, Cockroaches, & Rats* (Seattle: University of Washington Press, 2013).

49. R. Levine, e-mail message to author, February 27, 2013.

50. D. Friedenzohn, interviews with author, July 25, 2012; August 30, 2012.

51. S. Morrison and C. Winston, *The Economic Effects of Airline Deregulation* (Washington, DC: Brookings Institute, 1986).

52. US Department of State, "Open Skies Agreements," http://www.state.gov/e/eb/tra/ata/.

53. Research and Innovation Technology Administration, Bureau of Transportation Statistics, "U.S. Air Carrier Traffic and Capacity Summary by Service Class," http://www.transtats.bts.gov/Fields.asp?Table_ID=264.

54. Research and Innovation Technology Administration, Bureau of Transportation Statistics, "T-100 International Market (All Carriers)," http://www.transtats.bts.gov/Fields.asp?Table_ID=260.

55. US Census Bureau, "International Data Base World Population: 1950–2050," updated June 2011, http://www.census.gov/population/international/data/idb/worldpopgraph.php.

56. R. L. Usinger, *Monograph of Cimicidae* (Lanham, MD: Entomological Society of America, 1966).

57. C. Montes et al., "Maintenance of a Laboratory Colony of *Cimex lectularius* (Hemiptera: Cimicidae) Using an Artificial Feeding Technique," *Journal of Medical Entomology* 39, no. 4: 675–79 (2002).

58. M. Takano-Lee et al., "An Automated Feeding Apparatus for In Vitro Maintenance of the Human Head Louse, *Pediculus capitis* (Anoplura: Pediculidae)," *Journal of Medical Entomology* 40, no. 6 (2003): 795–99.

59. S. Jones, lab interview with author, January 15, 2013.

60. H. Harlan, e-mail messages to author, February 15, 2013; February 19, 2013.

61. D. Moore and D. Miller, "Laboratory Evaluations of Insecticide Product Efficacy for Control of *Cimex lectularius*," *Journal of Economic Entomology* 99, no. 6 (2006): 2080–86.

62. A. Romero, "Insecticide Resistance in the Bed Bug: A Factor in the Pest's Sudden Resurgence?" *Journal of Medical Entomology*, 44, no. 2 (2007): 175–78.

63. E. Vargo et al., "Genetic Analysis of Bed Bug Infestations and Populations," *Proceedings of the Seventh International Conference on Urban Pests*, São Paulo, Brazil, 2011.

64. C. Boase et al. "Interim Report on Insecticide Susceptibility Status of UK Bedbugs," *Professional Pest Controller*, Summer 2006.

65. C. Boase, interview with author, January 31, 2013.

66. D. Lilly et al., "Bed Bugs that Bite Back," *Professional Pest Manager*, August/ September 2009.

67. K. Yoon et al., "Biochemical and Molecular Analysis of Deltamethrin Resistance in the Common Bed Bug (Hemiptera: Cimicidae)," *Journal of Medical Entomology* 45, no. 6 (2008): 1092–101.

68. J. Clark, interview with author, January 30, 2013.

69. W. Brown, "Chapter 6: Resistance in the Housefly," *Insecticide Resistance in Arthropods* (World Health Organization, 1971).

70. J. Busvine, "The Significance of Insecticide-Resistant Strains," *Bulletin of the World Health Organization* 15 (1956): 389–401.

71. F. Zhu, "Widespread Distribution of Knockdown Resistance Mutations in the Bed Bug, *Cimex lectularius* (Hemiptera: Cimicidae), Populations in the United States," *Archives of Insect Biochemistry and Physiology* 73, no. 4 (2010): 245–57.

72. W. Booth, interviews with author, March 6, 2012; February 21, 2013; September 17, 2013; January 26, 2014.

73. C. Schal, interviews with author, October 30, 2012; June 18, 2013.

74. E. Vargo, interviews with author, October 30, 2012; February 7, 2013; February 15, 2013.

75. E. Vargo, e-mail message to author, January 6, 2014.

76. F. Zhu et al., "Bed Bugs Evolved Unique Adaptive Strategy to Resist Pyrethroid Insecticides," *Scientific Reports* 3 (2013): 1456.

77. W. Booth et al., "Molecular Markers Reveal Infestation Dynamics of the Bed Bug (Hemiptera: Cimicidae) within Apartment Buildings," *Journal of Medical Entomology* 49, no. 3 (2012): 535–46.

78. V. Saenz et al., "Genetic Analysis of Bed Bug Populations Reveals Small Propagule Size within Individual Infestations but High Genetic Diversity Across Infestations from the Eastern United States," *Journal of Medical Entomology* 49, no. 4 (2012): 865–75.

79. Interviews with members from the Essig Museum curatorial team, Ohio Bed Bug Task Force meeting, January 16, 2013.

80. M. Potter, interviews with author, January 28, 2013; February 12, 2013.

81. A. Coghlan, "Bedbugs Bite Back," *New Scientist*, October 5, 2002.

82. CDC, "CDC Malaria Map Application," http://www.cdc.gov/malaria/map/.

83. CDC, "Where Malaria Occurs," http://www.cdc.gov/malaria/about/distri bution.html.

84. S. Lindsay et al., "Permethrin-Impregnated Bednets Reduce Nuisance Arthropods in Gambian Houses," *Medical and Veterinary Entomology* 3, no. 4 (1989): 377–83.

85. E. Temu et al., "Bedbug Control by Permethrin-Impregnated Bednets in Tanzania," *Medical and Veterinary Entomology* 13, no. 4 (1999): 457–59.

86. J. Myamba et al., "Pyrethroid Resistance in Tropical Bedbugs, *Cimex hemipterus*, Associated with Use of Treated Bednets," *Medical Veterinary Entomology* 16, no. 4 (2002): 448–51.

Some of the passages of the chapter were adapted from pieces previously published by the author through other outlets:

1. Portions of the description on spiders appeared in "Slandered Spiders: That Probably Isn't a Brown Recluse Bite," *Slate*, October 28, 2013.
2. Portions of the description of artificial bed bug feeders appeared in "How Researchers Feed Bed Bugs in the Lab," PopularScience.com, February 5, 2013.
3. Portions of the description of genotyping appeared in "Using DNA to Track the Spread of Bedbugs," PopularScience.com, July 16, 2012.

CHAPTER FIVE

Sources for this chapter are listed below, in the order in which they appear. Sources used multiple times are listed only once.

1. J. Baker, J. "Apartment Fire Displaces 14," *Cincinnati Enquirer*, October 13, 2009.
2. M. Freese, interview with author, March 25, 2013.
3. C. Palm, interview with author, March 28, 2013.
4. Centers for Disease Control and Prevention, "Acute Illnesses Associated with Insecticides Used to Control Bed Bugs—Seven States, 2003–2010," *Morbidity and Mortality Weekly Report* 60, no. 37 (2011): 1269–74.
5. M. Beal, interviews with author, January 16, 2013; March 27, 2013.
6. Environmental Protection Agency, "Pesticide Registration Notice 97-1: Agency under the Requirements of the Food Quality Protection Act"; Common Mechanism Groups; EPA, Cumulative Exposure and Risk Assessment.
7. A. Gore, "Rachel Carson and *Silent Spring*," in *Courage for the Earth: Writers, Scientists, and Activists Celebrate the Life and Writing of Rachel Carson*, ed. Peter Matthiessen (New York: Houghton Mifflin, 2007).
8. A. Gore, introduction to *Silent Spring* by Rachel Carson (New York: Houghton Mifflin, 1994).
9. Environmental Protection Agency, interview with author, January 29, 2014.
10. Environmental Protection Agency, "The Food Quality Protection Act (FQPA) Background," 1996.
11. Environmental Protection Agency, "Pesticide Registration (PR) Notice 91-1: Agency Actions under the Requirements of the Food Quality Protection Act," January 31, 1997.
12. Environmental Protection Agency, "Pesticide Fact Sheet: Chlorfenapyr."
13. Environmental Protection Agency, "Pesticide Fact Sheet: Insect Growth Regulators: S-Hydroprene (128966), S-Kinoprene (107502), Methoprene (105401), S-Methoprene (105402)."
14. A. Romero and M. Potter, "Evaluation of Chlorfenapyr for Control of the Bed Bug, *Cimex lectularius* L.," *Pest Management Science* 66, no. 11 (November 2010): 1243–48.
15. M. Potter, "Bed Bug Insecticides, Seeking a Silver Bullet" (research presented at the BedBug University North American Summit, Las Vegas, NV, September 6–7, 2012).
16. M. Potter et al., "Dual-Action Bed Bug Killers," *PCT Online*, March 2012.
17. M. Potter, interview with author, December 16, 2013.

18. M. Goodman, "Effects of Juvenile Hormone Analog Formulations on Development and Reproduction in the Bed Bug *Cimex lectularius* (Hemiptera: Cimicidae)," *Pest Management Science* 69, no. 2 (2013): 240–44.

19. A. Rosenfeld, "The Ultimate Weapon in an Ancient War," *LIFE*, October 6, 1958.

20. V. Wigglesworth, "Memoirs: The Physiology of Ecdysis in *Rhodnius Prolixus* (Hemiptera). II. Factors Controlling Moulting and 'Metamorphosis,'" *Quarterly Journal of Microscopical Science*, s2-77 (1934): 191–222.

21. V. Wigglesworth, "The Function of the Corpus Allatum in the Growth and Reproduction of *Rhodnius prolixus* (Hemiptera)," *Quarterly Journal of Microscopic Science* 79 (1936): 91–121.

22. M. Potter, "The History of Bed bug Management," *American Entomologist* 57 (2011).

23. G. Fox, *The Mediaeval Sciences in the Works of John Gower* (New York: Haskell House, 1966).

24. Pliny the Elder, *The Natural History*, trans. John Bostock (London: Taylor and Francis, Red Lion Court, Fleet Street, 1855).

25. Ohio State University, *Thirty-Sixth Annual Report of the Board of Trustees of The Ohio State University to the Governor of Ohio for the Year Ending June 30, 1906.*

26. V. Bogdandy, *Die Naturwissenshaften* 48 (1927): 474.

27. K. Dewhurst, *John Locke, 1632–1704, Physician and Philosopher: A Medical Biography* (London: Wellcome Historical Medical Library, 1963).

28. M. de Scaevola, "Archives of Poitou on Bed Bugs, Letter from Secretary of the King," Argenton Berry, M. Jouineau Desloges. February 22, 1777.

29. US Patent 8,883; US Patent 391,930; US Patent 589,465; US Patent 421,604; US Patent 132,080; US Patent 481,270.

30. J. Charnley, United States Patent and Trademark Office, e-mail message to author, November 19, 2012.

31. M. Potter, "The History of Bed Bug Management—With Lessons from the Past," *American Entomologist*, Spring 2011.

32. J. Southall, *Treatise on the Cimex lectularius; or bed bug* (London: J. Roberts, 1730).

33. J. Southall, *Treatise on the Cimex lectularius; or bed bug. The second edition with additions, by a physician* (British Library, 1783; reprint, Farminghills, MI: Gale ECCO, 2010).

34. *Proceedings of the Eleventh Annual Convention of the State Association of Superintendents of Poor and Keepers of County Infirmaries* (Lansing, MI: Wynkoop Hallenbeck Crawford Co. State Printers, 1915).

35. P. Stillwell, *Battleship Arizona: An Illustrated History* (Annapolis, MD: US Naval Academy Press, 1991).

36. Ohio State University, *Thirty-Sixth Annual Report of the Board of Trustees of The Ohio State University to the Governor of Ohio for the Year Ending June 30, 1906.*

37. J. R. Wells, *A new and valuable book, entitled the Family Companion, Containing Many Hundred Rare and Useful Receipts on Every Branch of Domestic Economy* (Boston: Printed for the Author, 1846).

38. S. Ashmore and A. McKenney, "Coal-Tar Naphtha Distillates for Destruction of Bed-Bugs," *British Medical Journal* 459 (1937).

39. P. Stock, "The Utilization of Lethal Gases in Hygiene," *Proceedings of the Royal Society of Medicine* 31 (1938): 427–42.

40. R. Creel, "Cyanide Gas for the Destruction of Insects," *Public Health Reports* 31 (1916): 1464–75.

41. R. Corca, "WWII Army Barracks Disinfestation Photos," *New York vs. Bed Bugs*, http://newyorkvsbedbugs.org/2010/08/01/wwii-army-barracks-disinfestation-photos/ (photographs confirmed by other sources).

42. L. McMurry, *George Washington Carver: Scientist and Symbol* (New York: Oxford University Press, 1981).

43. P. McDougall, *The Cost of New Agrochemical Product Discovery, Development and Registration in 1995, 2000, and 2005–2008 R&D Expenditure in 2007 and Expectations for 2012 Final Report*, Consultancy Study for Crop Life America and the European Crop Protection Association, January 2010.

44. B. Reid, Bayer senior principal scientist, interview with author, April 4, 2013.

45. J. Cink, BASF global project manager, interview with author, April 3, 2013.

46. C. Scherer, Syngenta product development, interview with author, April 1, 2013.

47. V. Gutsmann, Bayer CropScience product development, interviews with author, May 23–24, 2013.

48. M. Adamczewski, Bayer CropScience research scientist, interview with author, May 24, 2013.

49. S. Horstmann, Bayer CropScience product development, interviews with author, May 23–24, 2013.

50. C. Scherer, e-mail message to author, April 23, 2013.

51. D. Leva, of DuPont, e-mail message to author, April 23, 2013.

52. G. Georghiou, "Overview of Insecticide Resistance," *ACS Symposium Series* 421 (1990).

53. D. Pimentel, e-mail message to author, May 16, 2012.

54. Environmental Protection Agency, "Minimum Risk Pesticides," last updated January 3, 2013, http://www.epa.gov/oppbppd1/biopesticides/regtools/25b_list.htm.

55. Environmental Protection Agency, "Federal Insecticide, Fungicide, and Rodenticide Act (FIFRA)," last updated June 27, 2012, http://www.epa.gov/oecaagct/lfra.html.

56. Environmental Protection Agency, interview with author, January 31, 2013.

57. H. Richardson, "The Action of Bean Leaves Against the Bedbug," *Journal of Economic Entomology* 36, no. 4 (1943): 543–45.

58. M. Szyndler, "Entrapment of Bed Bugs by Leaf Trichomes Inspires Microfabrication of Biometic Surfaces," *Journal of the Royal Society Interface* 10, no. 83 (2013).

59. "Innovative New Nanotechnology Stops Bed Bugs in Their Tracks—Literally," Stony Brook University Press Release, May 30, 2013.

60. J. White, interviews with author, June 6, 2011; September 20, 2012.

61. P. Cooper, interviews with author, September 6, 2012; September 7, 2012; September 20, 2012.

62. S. Jones, interviews with author, November 20, 2012; January 15, 2013; January 16, 2013; January 25, 2013.

63. J. Logan and M. Cameron, interview with author, May 8, 2013.

64. J. Benoit et al., "Addition of Alarm Pheromone Components Improves the Effectiveness of Desiccant Dusts Against *Cimex lectularius*," *Medical Entomology* 46, no. 3 (2009): 572–79.

65. V. Harraca et al., "Nymphs of the Common Bed Bug (*Cimex lectularius*) Produce Anti-Aphrodisiac Defence against Conspecifc Males," *BMC Biology* 8, no. 117 (2010).

66. K. Haynes et al., "Bed Bug Deterrence," *BMC Biology* 8, no. 117 (2010).

67. Environmental Protection Agency, "What Are Biopesticides?," http://www.epa.gov/pesticides/biopesticides/whatarebiopesticides.htm.

68. A. Barbarin et al., "A Preliminary Evaluation of the Potential of *Beauveria bassiana* for Bed Bug Control," *Journal of Invertebrate Pathology* 111 (2012): 82–85.

69. N. Jenkins, interview with author, April 25, 2013.

70. A. González Canga et al., "The Pharmacokinetics and Interactions of Ivermectin in Humans: A Mini-Review," *AAPS Journal* 10, no. 1 (2008): 42–46.

71. M. Hertig and S. Wolbach. "Studies on Rickettsia-Like Micro-Organisms in Insects," *Journal of Medical Research* 44, no. 3 (1924): 329–74.

72. A. Hoerauf and R. Rao, *Wolbachia: A Bug's Life in Another Bug* (Basel, Switzerland: Karger, 2007).

73. K. Hilgenboecker et al., "How Many Species Are Infected with *Wolbachia*?: A Statistical Analysis of Current Data," *FEMS Microbiology Letters* 281, no. 2 (2008): 215–20.

74. T. Hosokawa et al., "*Wolbachia* as a Bacteriocyte-Associated Nutritional Mutualist," *PNAS* 107, no. 2 (2010): 769–74.

Some of the passages of the chapter were adapted from pieces previously published by the author through other outlets:

1. Portions of the description of the bean leaves experiments appeared in "Building a Better Bed Bug Trap," PopularScience.com, April 9, 2013.

2. Portions of the description of the ivermectin experiment appeared in "Can Taking a Pill before Bed Get Rid of Bed Bugs?," PopularScience.com, November 26, 2012.

CHAPTER SIX
Sources for this chapter are listed below, in the order in which they appear. Sources used multiple times are listed only once.

1. M. Antony and D. Barlow, "Specific Phobias," in D. Barlow, *Anxiety and Its Disorders* (New York: Guilford Press, 2002).

2. J. Lockwood, *The Infested Mind: Why Humans Fear, Loathe, and Love Insects* (New York: Oxford University Press, 2013).

3. P. McNamara, e-mail message to author, July 30, 2013.

4. A. R. Ekirch, *At Day's Close: Night in Times Past* (New York: Norton, 2005).
5. A. R. Ekirch, interview with author, August 14, 2013.
6. T. Cole, *Open City* (New York: Random House, 2011).
7. M. Lehane, *The Biology of Blood-Sucking in Insects*, 2nd ed. (Cambridge: Cambridge University Press, 2005).
8. M. Lehane, "Blood Sucking," *Encyclopedia of Insects*, 2nd ed. (Burlington, MA: Academic Press, 2009).
9. P. Delaunay et al., "Bedbugs and Infectious Diseases," *Clinical Infectious Diseases* 52, no. 2 (2011): 200–210.
10. W. Rucker, "The Bedbug," *Public Health Reports (1896–1970)*, 27, no. 46 (1912): 1854–56.
11. J. Hutchison, "Leprosy and the Cimex," *British Medical Journal*, August 26, 1911.
12. C. Asher, *Bacteria Inc.* (Boston: Bruce Humphries, 1995).
13. "Seventy-Third Annual Meeting of the British Medical Association," *British Medical Journal*, November 1905.
14. J. Dodds Price and L. Rogers, "The Uniform Success of Segregation Measures in Eradicating Kala-Azar from Assam Tea Gardens," *British Medical Journal*, February 7, 1914.
15. H. Durham, "Notes on Beriberi in the Malay Peninsula and on Christmas Island (Indian Ocean)," *Journal of Hygiene* IV (1904).
16. D. Bergey, *The Principles of Hygiene* (Philadelphia: W. B. Saunders Company, 1909).
17. D. Thomson, "Preliminary Note on Bed-Bugs and Leprosy," *British Medical Journal*, October 4, 1913.
18. W. Riley and O. Johannsen, *Handbook of Medical Entomology* (Ithaca, NY: Comstock Publishing, 1915).
19. G. Burton, "Bedbugs in Relation to Transmission of Human Diseases," *Public Health Reports* 78, no. 6 (1963): 513–24.
20. "Scientific Proceedings of the Twenty-Third Annual Meeting of the American Association of Pathologists and Bacteriologists," Boston, MA, March 30–31 1923.
21. J. Arkwright, "President's Address: The Unity of Medicine," *Proceedings of the Royal Society of Medicine*, October 28, 1931.
22. M. Ruiz Castaneda and H. Zinsser, "Studies on Typhus Fever: Studies of Lice and Bedbugs (*Cimex lectularius*) with Mexican Typhus Fever Virus," *Journal of Experimental Medicine* 52, no. 5 (1930): 661–68.
23. N. Rogers, "Dirt, Flies, and Immigrants: Explaining the Epidemiology of Poliomyelitis, 1900–1916," *Sickness and Health in America* (Madison: University of Wisconsin Press, 1978).
24. C. Campbell, "My Observations on Bedbugs," *Bats, Mosquitoes and Dollars* (Boston: Stratford Company, 1925).
25. R. L. Usinger, *Monograph of Cimicidae* (Lanham, MD: Entomological Society of America, 1966).
26. W. Wills et al., "Hepatitis-B Virus in Bedbugs (*Cimex hemipterus*) from Senegal," *Lancet* 2 (1977): 217–19.
27. "Bed Bugs, Insects and Hepatitis B," *British Medical Journal*, September 29, 1979.

28. B. Brotman et al., "Role of Arthropods in Transmission of Hepatitis-B Virus in the Tropics," *Lancet* 1 (1973): 1305–8.

29. M. Mayans et al., "Do Bedbugs Transmit Hepatitis B?" *Lancet* 343 (1994): 761–63.

30. P. Jupp et al., "Attempts to Transmit Hepatitis B Virus to Chimpanzees by Arthropods." *South African Medical Journal* 79 (1991): 320–22.

31. P. Jupp and S. McElligott, "Transmission Experiments with Hepatitis B Surface Antigen and the Common Bedbug (*Cimex lectularius L.*)," *SA Medical Journal* 56 (1979): 54–57.

32. R. Gallo et al., "Isolation of Human T-Cell Leukemia Virus in Acquired Immune Deficiency Syndrome (AIDS)," *Science* 220, no. 4599 (1983): 865–67.

33. F. Barre-Sinoussi et al., "Isolation of a T-Lymphotropic Retrovirus from a Patient at Risk for Acquired Immune Deficiency Syndrome (AIDS)," *Science* 220, no. 4599 (1983): 868–71.

34. P. Jupp and S. Lyons, "Experimental Assessment of Bedbugs (*Cimex lectularius* and *Cimex hemipterus*) and Mosquitoes (*Aedes aegypti formosus*) as Vectors of Human Immunodeficiency Virus," *AIDS* 1, no. 3 (1987): 171–74.

35. P. Webb, "Potential for Insect Transmission of HIV: Experimental Exposure of *Cimex hemipterus* and *Toxorynchites amboinensis* to Human Immunodeficiency Virus," *Journal of Infectious Disease* 160, no. 6 (1989): 970–77.

36. A. Zuckerman, "AIDS and Insects," *British Medical Journal* 292 (1986): 1094–95.

37. E. Vargo and C. Schal, interview with author, November 30, 2012.

38. V. Saenz et al., "Survey of *Bartonella* spp. in U.S. Bed Bugs Detects *Burkholderia multivorans* but Not *Bartonella*," *PLoS ONE* 8, no. 9 (2013): e73661

39. C. Darrington and S. Jones, interview with author, January 15, 2013; July 10, 2013.

40. C. Lowe and M. Romney, "Bedbugs as Vectors for Drug-Resistant Bacteria [letter]," *Emerging Infectious Diseases*, June 2011.

41. Centers for Disease Control and Prevention, "Vancomycin-Resistant Enterococci in Healthcare Settings," http://www.cdc.gov/hai/organisms/vre/vre.html.

42. Centers for Disease Control and Prevention, "MRSA Infections," http://www.cdc.gov/mrsa/definition/index.html.

43. S. Doggett et al., "Bed Bugs: Clinical Relevance and Control Options," *Clinical Microbiology Reviews* 25, no. 1 (2012): 164–92.

44. A. Dobson, interview with author, July 15, 2013.

45. J. Peterson, e-mail messages to author, July 12, 2013; December 16, 2013; January 2, 2014.

46. W. Booth, e-mail message to author, October 11, 2012.

47. C. Brown et al., "Natural Infection of Vertebrate Hosts by Different Lineages of Buggy Creek Virus (family Togaviridae, genus *Alphavirus*)," *Archives of Virology* 155, no. 5 (2010): 745–49.

48. A. Teixeira et al., "Evolution and Pathology in Chagas Disease: A Review," *Memórias do Instituto Oswaldo Cruz* 101, no. 5 (2006): 463–91.

49. J. Goddard and R. deShazo, "Bed Bugs (*Cimex lectularius*) and Clinical Consequences of Their Bites," *Journal of the American Medical Association* 301, no. 12 (2009): 1358–66.

50. Z. Adelman et al., "Bed Bugs and Infectious Disease: A Case for the Arboviruses," *PLOS Pathogens* 9, no. 8 (August 2013).
51. Z. Adelman, interview with author, August 26, 2013.
52. H. Stelwagon, *Treatise on Diseases of the Skin* (Philadelphia: W. B. Saunders Company, 1910).
53. S. Doggett and R. Russell, "Bed Bugs: What the GP Needs to Know," *Australian Family Physician* 38, no. 11 (2009): 880–84.
54. R. deShazo et al., "Bullous Reactions to Bedbug Bites Reflect Cutaneous Vasculitis," *American Journal of Medicine* 125, no. 7 (2012): 688–94.
55. M. Feldlaufer, interview with author, November 30, 2012.
56. M. Sabou et al., "Bed Bugs Reproductive Life Cycle in the Clothes of a Patient Suffering from Alzheimer's Disease Results in Iron Deficiency Anemia," *Parasite* 20, no. 16 (2013): 1–5.
57. M. Paulke-Korinek et al., "Bed Bugs Can Cause Severe Anaemia in Adults," *Parasitology Research* 110, no. 6 (2012): 2577–79.
58. M. J. Pritchard and S. Hwang, "Severe Anemia from Bedbugs," *Canadian Medical Association Journal* 181, no. 5 (2009): 287–88.
59. H. Harlan, interview with author, July 8, 2011.
60. J. Goddard and R. deShazo, "Psychological Effects of Bed Bug Attacks (*Cimex lectularius* L.)," *American Journal of Medicine* 125, no. 1 (2012) 101–3.
61. H. Colten and B. Altevogt, *Sleep Disorders and Sleep Deprivation* (Washington, DC: National Academies Press, 2006).
62. National Highway Traffic Safety Administration, "Research on Drowsy Driving," http://www.nhtsa.gov/Driving+Safety/Distracted+Driving+at+Distraction.gov/Research+on+Drowsy+Driving.
63. United States Nuclear Regulatory Commission, "Report on the Accident at the Chernobyl Nuclear Power Station," January 1987.
64. Alaska Oil Spill Commission, "Spill: The Wreck of the *Exxon Valdez*," February 1990.
65. S. Susser et al., "Mental Health Effects from Urban Bed Bug Infestation (*Cimex lectularius* L.): A Cross-Sectional Study," *BMJ Open* 2 (2012): e000838.
66. S. Perron, interview with author, October 3, 2011.
67. E. Rieder et al. "Psychiatric Consequences of Actual Versus Feared and Perceived Bed Bug Infestations: A Case Series Examining a Current Epidemic," *Psychosomatics* 53 (2012): 85–91.
68. E. Rieder, interview with author, August 1, 2013.
69. N. Hinkle, "Delusory Parasitosis," *American Entomologist* 45, no. 1 (2000): 17–25.
70. N. Hinkle, "Ekbom Syndrome: The Challenge of 'Invisible Bug' Infestations," *Annual Review of Entomology* 55 (2010): 77–94.
71. "Valid and Putative Human Skin Infestations" (presentations at the National Conference of Urban Entomology in Atlanta, GA, May 22, 2012).
72. S. Dalí, S. *The Secret Life of Salvador Dalí* (Mineola, NY: Dover, 1993).
73. S. Burrows et al., "Suicide Following an Infestation of Bed Bugs," *American Journal of Case Reports* 14 (2013): 176–78.

Some of the passages of the chapter were adapted from pieces previously published by the author:

1. Portions of the description of the evolution of bloodsucking insects appeared in "How Did Bloodsucking Insects Evolve?," *Our Modern Plagues*, PopularScience.com, January 16, 2014.
2. Portions of the description of Salvador Dalí's delusory parasitosis appeared in "Was Salvador Dalí Plagued by Phantom Bed Bugs?," *Our Modern Plagues* PopularScience.com, October 29, 2013.

CHAPTER SEVEN

Sources for this chapter are listed below, in the order in which they appear. Sources used multiple times are listed only once.

1. K. Reinhardt et al., "Situation Exploitation: Higher Male Mating Success When Female Resistance Is Reduced by Feeding," *Evolution* 63, no. 1 (January 2008): 29–39.
2. K. Reinhardt et al., "Potential Sexual Transmission of Environmental Microbes in a Traumatically Inseminated Insect," *Ecological Entomology* 30 (2005): 607–11.
3. M. Siva-Jothy, "Know Your Enemy: Why Pure Research in Relevant for Bed Bug Control" (presentation at BedBug University Summit, Las Vegas, NV, September 6–7, 2012).
4. M. Seva, "Bedbugs in the Duvet," *New York Magazine*, May 2, 2010.
5. P. Cooper, interviews with author September 20, 2012; December 6, 2012; and subsequent e-mail messages.
6. B. Hirsch, e-mail message to author, December 19, 2013.
7. M. Eisemann, interviews with author, September 6, 2012; November 27, 2012.
8. M. Eisemann, e-mail message to author, December 9, 2013.
9. Amazon.com search for "bed bug" ("bedbug" search recommends spelling "bed bug" and Google Shopping search for "bed bug" OR bedbug (October 1, 2013).
10. R. Naylor, e-mail message to author, April 27, 2013.
11. C. Schal and E. Vargo, interview with author, October 30, 2012; Bill Donahue, e-mail message to author, January 4, 2013.
12. "CimexScent Product to Become Available," i2LResearch Ltd., http://www.icrlab.com/news/.
13. i2LResearch USA, e-mail messages to author, November 2013.
14. "A Strategic Analysis of the U.S. Structural Pest Control Industry: The 2011 Season" (published April 2012).
15. G. Curl, e-mail messages to author, September 20, 2012; December 8, 2012; January 12, 2013; October 1, 2013.
16. M. Potter et al., "The 2013 Bugs without Borders Survey" (National Pest Management Association, 2013).
17. M. Potter et al., "Bugs without Borders: Defining the Global Bed Bug Resurgence" (National Pest Management Association, 2010).
18. Unpublished data by author of patents and published patent applications covering entire history of patents through December 31, 2012, using CobaltIP patent analyzing software, www.freepatentsonline.com, and

Google Patents. Search terms through all fields: "bed bug" OR bedbug OR bed-bug OR "bed bugs" OR bedbugs OR bed-bugs OR "Cimex lectularius," across all available patent databases worldwide. Additional searches for bed bug classification (class 43, subclass 123) and historical terms such as "chinches." Irrelevant patents removed from final database.

19. *Nobel Lectures in Physiology or Medicine: 1942–1962* (Singapore: World Scientific Publishing Co. Pte. Ltd., 1999).

20. M. Curcio, "Bug Housing for Attracting, Monitoring, and Detecting Bugs," United States Patent Published Application 2012/0291337 A1.

21. M. Curcio, "Bug Housing for Attracting, Monitoring, and Detecting Bugs," World Patent 2012/158140 A1.

22. "Friday Final TV Ratings: 'Shark Tank' Adjusted Up, 'America's Next Top Model' Adjusted Down," *TV by the Numbers*, September 17, 2012, http://tv bythenumbers.zap2it.com/2012/09/17/friday-final-tv-ratings-shark-tank -adjusted-up-americas-next-top-model-adjusted-down/148821/.

23. Shark Tank, "Week 1," *Shark Tank* video 31:52–42:11, September 14, 2012, https://www.youtube.com/watch?v=3rnBFYaWuns and http://vimeo.com /50405770.

24. M. Curcio, interview with author, December 9, 2012.

25. M. Curcio, e-mail message to author, December 12, 2012.

26. M. Smith, of ABC, e-mail message to author, December, 12, 2012.

27. *Svetlana Tendler and Jacob Tendler v. Hilton Worldwide Inc. d/b/a Waldorf Astoria Hotels & Resorts, The Blackstone Group, L.P.*, Supreme Court of the State of New York, New York County Index No. 110704/2010.

28. "New York's Waldorf Astoria that Heralds Itself as Embodiment of Luxury and Splendor Sued for 10 Million for Bed Bug Infestation," PR WEB, January 6, 2011, http://www.prweb.com/releases/2011/bedbugsatwaldorfastoria /prweb4943854.htm.

29. *Christine Drabicki and David Drabicki v. The Waldorf-Astoria Hotel*, Supreme Court of the State of New York, New York County Index No. 10114455.

30. C. Hoffberger, "Hotel Sues Man over Lousy Review," *Salon*, August 8, 2013, http://www.salon.com/2013/08/24/hotel_sues_man_95000_for_a_bad _trip_advisor_review_partner/?source=newsletter.

31. J. Lipman, interviews and e-mail messages to author September 2012– February 2013.

32. AOL Real Estate Editors, "Tenant Faika Shaaban Awarded $800,000 for Bedbug Infestation," *AOL*, June 4, 2013 http://realestate.aol.com/blog/on /faika-shaaban-wins-annapolis-bedbug-lawsuit/.

33. "Man Slept Tight, but the Bedbugs Bite Anyway," *Jet*, July 31, 1975.

34. *Lawrence M. Delamater v. Sarah Foreman and Another*, 239 N.W. 148, 149 (Minn. 1931).

35. M. Berenbaum, *Bugs in the System: Insects and Their Impact on Human Affairs* (New York: Basic Books, 1995).

36. B. Rich et al., *American Law Reports Annotated*, vol. 4 (Rochester: Lawyers Co-Operative Publishing Company, 1919).

37. K. Sweeney, "Pesticide Efficacy Testing Guidelines, 25b Products and the Truth About Advertising Laws" (presentation at BedBug University Summit, Las Vegas, NV, September 6–7, 2012).

38. S. Jennings, "EPA's Role in Bed Bug Products Exempt from FIFRA" (presentation at BedBug University Summit, Las Vegas, NV, September 6–7, 2012).

39. K. O'Brien, "Recent FTC Initiatives" (presentation at BedBug University Summit, Las Vegas, NV, September 6–7, 2012).

40. K. O'Brien, interviews with author, September 17, 2012, and January 17, 2013.

CHAPTER EIGHT

Sources for this chapter are listed below, in the order in which they appear. Sources used multiple times are listed only once.

1. D. Cain, interview with author, May 10, 2013.

2. D. Cain, e-mail message to author, December 17, 2013.

3. J. Logan, interview with author, May 8, 2013.

4. M. Cameron, interview with author, May 8, 2013.

5. A. Robinson, interview with author, May 8, 2013.

6. M. Siva-Jothy, interview with author, May 9, 2013.

7. K. Reinhardt, interview with author, November 8, 2012.

8. K. Reinhardt, e-mail message to author, December 17, 2013.

9. R. Naylor, e-mail message to author, December 17, 2013.

10. M. Siva-Jothy, personal field notes, 2004.

11. M. Siva-Jothy, personal photo collection.

12. K. Reinhardt et al., "Temperature and Humidity Differences between Roosts of the Fruit Bat, *Rousettus aegyptiacus* (Geoffroy, 1810), and the Refugia of Its Ectoparasite, *Afrocimex constrictus*," *Acta Chiropterologica* 10, no. 1 (2008): 173–76.

13. K. Reinhardt et al., "Estimating the Mean Abundance and Feeding Rate of a Temporal Ectoparasite in the Wild: *Afrocimex constrictus* (Heteroptera: Cimicidae)," *International Journal of Parasitology* 37 (2007): 937–42.

14. S. Wynne-Jones and M. Walsh, "Heritage, Tourism, and Slavery at Shimoni: Narrative and Metanarrative on the East African Coast," *History in Africa* 37 (2010): 247–73.

15. M. Tuttle, "The Lives of Mexican Free-Tailed Bats," *Bats Magazine* 12, no. 3 (Fall 1994).

16. K. Reinhardt and S. Roth, "Protecting Bat Caves Conserves Diverse Ecosystems," *Bats Magazine* 31, no. 3 (2013): 2–4.

CHAPTER NINE

Sources for this chapter are listed below, in the order in which they appear. Sources used multiple times are listed only once.

1. H. Sounes, *Charles Bukowski: Locked in the Arms of a Crazy Life* (New York: Grove Press, 1998).

2. O. Balvin, interviews with author, May 12–22, 2013.

3. R. L. Usinger, *Robert Leslie Usinger: Autobiography of an Entomologist* (San Francisco: Pacific Coast Entomological Society, 1972).

4. M. George, interview with Dalibor Povolný, *Messenger* 4, no. 1 (2004): 5–12.

5. Various correspondence from 1957 to 1968, Povolný folder, box 2, Usinger Papers, Bancroft Library, University of California, Berkeley.

6. Federal Bureau of Investigation File 123247, 1963, Dalibor Povolný.
7. Federal Bureau of Investigation File 65-30092-5094, 1954, Robert Usinger.
8. "Visiting Professor Studies Parasites Carrying Diseases," *Kabul Times*, May 1, 1966.
9. R. L. Usinger and D. Povolný, "The Discovery of a Possibly Aboriginal Population of the Bed Bug (*Cimex lectularius* Linnaeus, 1758)," *Acta Musei Moravie* (1966).
10. F. Rettich, interview with author, May 14, 2013.
11. F. Rettich, e-mail messages to author, January 5, 2014.
12. Z. Galková, interview with author, May 14, 2013.
13. L. Kučerová, interview with author, May 14, 2013.
14. H. Smíšková, interview with author, May 20, 2013.
15. J. Šembera, interview with author, May 21, 2013.
16. J. Višnička, interview with author, May 22, 2103.
17. Nováčany and Luník 9 interviews, May, 2013.
18. A. Higgins, "In Its Efforts to Integrate Roma, Slovakia Recalls U.S. Struggles," *New York Times*, May 9, 2013.
19. M. Bačo, interview with author, May 16, 2013.
20. M. Bačo, e-mail message to author, January 5, 2014.
21. S. Anthony, interviews with author, September 17, 2013, and January 20, 2014.
22. S. Doggett et al., "Bed Bugs: Clinical Relevance and Control Options," *Clinical Microbiology Reviews* 25, no. 1 (2012): 164–92.
23. O. Balvin, e-mail message to author, September 20, 2013.
24. K. Wawrocka and T. Bartonička, "Two Different Lineages of Bedbug (*Cimex lectularius*) Reflected in Host Specificity," *Parasitology Research*, August 28. 2013.
25. W. Booth, interview with author, September 17, 2013.
26. W. Booth et al., "Host Association Drives Deep Divergence in the Common Bed Bug, *Cimex lectularius*" (unpublished manuscript, 2014).

EPILOGUE

Sources for this chapter are listed below, in the order in which they appear. Sources used multiple times are listed only once. I also owe thanks to Jeffrey Lockwood, whose book *The Infested Mind* pointed me both to the William Gass short story and to Karl von Frisch's *Ten Little Housemates*.

1. T. LeClair, "Interviews: William Gass, The Art of Fiction No. 65," *Paris Review*, http://www.theparisreview.org/interviews/3576/the-art-of-fiction-no -65-william-gass.
2. W. Gass, *In the Heart of the Heart of the Country & Other Stories* (New York: Harper & Row, 1968)
3. K. von Frisch, *Ten Little Housemates* (Oxford: Pergamon Press, 1960).

INDEX

Page numbers in italics refer to illustrations.